Paper Industry Waste Processing

This technical resource delves into the management and valorization of paper industry waste, exploring innovative approaches to transform waste into sustainable construction materials, low-grade paperboard, biogas, and bioCNG. It provides a comprehensive overview of cutting-edge advancements in recycling processes, recovery technologies, and the production of clean hydrogen from paper mill waste. By bridging theoretical research with practical industrial applications, the book systematically presents core principles of recovery technologies alongside real-world case studies. Additionally, it emphasizes the role of waste management in fostering a circular economy within the pulping sector, promoting environmental sustainability and resource efficiency.

- Examines fundamental issues relating to industrial pulp packaging
- Includes theory of industry waste conversion into eco-friendly packaging materials
- Discusses comprehensive cost analysis of industrial pulp packaging with plastic/thermocol-based packaging
- Reviews available technology for paper mill and forest products industry wastes in connection with sustainability concepts
- Explores trends on value-added products from biorefinery lignin and reducing black liquor load

This book is aimed at graduate students and researchers in pulp and paper technology, waste management, and chemical process engineering.

Paper Industry Waste Processing

A Circular Economy Approach

Suresh Sundaramurthy, Saakshy, Anuj Kumar Verma,
and Arisutha Suresh

CRC Press
Taylor & Francis Group
Boca Raton London New York

CRC Press is an imprint of the
Taylor & Francis Group, an **informa** business

Designed cover image: Shutterstock

First edition published 2026
by CRC Press
2385 NW Executive Center Drive, Suite 320, Boca Raton FL 33431

and by CRC Press
4 Park Square, Milton Park, Abingdon, Oxon, OX14 4RN

CRC Press is an imprint of Taylor & Francis Group, LLC

ISBN: 978-1-032-58926-8 (hbk)
ISBN: 978-1-032-60984-3 (pbk)
ISBN: 978-1-003-46139-5 (ebk)

DOI: 10.1201/9781003461395

Typeset in Times
by Apex CoVantage, LLC

Contents

Foreword

This document explores the integration of *circular economy principles in the paper industry,* a concept that promotes the reuse, recycling, and regeneration of materials to extend their lifecycle and reduce waste. By rethinking product design, production processes, and consumption models, the paper industry can evolve from a linear model of "take, make, dispose" to one that is regenerative by design. This work serves as both a critical examination and a forward-looking vision of how circular strategies can be effectively applied in the paper sector. It is my hope that the insights contained within will inform, inspire, and catalyze further progress toward a greener, circular future.

This book consists of seven chapters that highlight different objectives of a circular economy approach in the paper industry. These include waste utilization, characterization, as well as the circular economy and green opportunities for pulp and paper industry.

In Chapter 1, a current global overview of the pulp and paper industry is presented, focusing on country-wise export, import, consumption, and production of paper and paperboard in 2022 and 2023, to provide a better understanding of the significance of the circular economy in the paper industry.

Chapter 2 provides details on paper mill sludge utilization in fiber recovery, filler recovery, and low-grade paperboard production. Paper mill sludge represents one of the most significant waste streams in the pulp and paper industry, posing both environmental challenges and economic opportunities. This chapter explores the systematic approaches to transforming this industrial byproduct from an environmental liability into a valuable resource within the framework of circular economy principles. Through detailed examination of fiber and filler recovery techniques, applications in low-grade paperboard manufacturing, and sustainability perspectives, this chapter provides a comprehensive roadmap for paper industry stakeholders seeking to implement waste valorization strategies.

Chapter 3 provides details on the valorization aspects of paper mill sludge for the construction industry. Paper mill sludge can be utilized in the construction industry as a sustainable and cost-effective alternative to traditional building materials.

The utilization of paper mill sludge waste also adds to the circular economy of the paper industry, improves sustainability, reduces the environmental impact of construction by minimizing landfill disposal, and promotes resource efficiency.

Chapter 4 discusses the utilization of coal fly ash as a pigment and filler for papermaking. This chapter describes the usage of fly ash as a filler and pigment in the paper industry. In addition, different process parameters, such as optical properties, sheet formation, paper dimensional stability, printability, and appearance, are explained. For writing and printing paper, the result shows that fly ash-based filler has unique properties compared to PCC due to high-pressure carbonation and pre-flocculation techniques. Modified Fly Ash-Based Calcium Silicate (FACS) has high first-pass retention (FPR) and first-pass ash retention (FPAR) compared to unmodified Precipitated Calcium Carbonate (PCC) with lower starch consumption. Modified FACS shows a slight drop in opacity and brightness compared to unmodified PCC, as the treated filler has a very small negative impact on optical properties.

Chapter 5 covers decarbonization and the energy transition in waste management of the pulp and paper industry. The waste feedstocks can be transformed into an energy source, such as methane. This methane production enhances national energy security concerns, helps to decarbonize the paper production process, and also reduces the overload on conventional sources. Chapter 5 gives insight into the energy transition toward methane production and how it can be materialized for paper and pulp industrial waste management to minimize their impact on the environment.

Chapter 6 covers the role of the circular economy in mitigating the environmental impacts of the paper industry. This chapter covers the various aspects of the circular economy, including the life cycle of paper and the environmental aspects of paper recycling, for a better understanding of the circular economy. A review of existing studies emphasizes the importance of waste paper, wastewater, paper sludge, and other byproducts generated by the paper industry. Recycling of paper results in a decrease in the cutting of trees and deforestation and improves ecological sustainability. The pulp and paper industry is moving toward a circular economy for sustainable production.

Chapter 7 discusses the utilization of fly ash-based fillers and aims to demonstrate their preparation for papermaking. FPR and FPAR of fly ash-based fillers are very high compared to those of conventional fillers like PCC, talc, and Ground Calcium Carbonate (GCC) under identical conditions with and without retention aid. FACS fillers have high sizing demand compared to conventional fillers; on increasing the sizing dosages, FACS filler sheets can be properly sized but at the expense of high sizing chemical consumption. A blend of FACS and a conventional filler can provide a solution to this problem.

Circular economy presents not only an environmental imperative but also an economic opportunity. It encourages collaboration across the value chain, from forestry and pulp production to recycling facilities and end consumers. Through innovation, policy support, and stakeholder engagement, the paper industry can become a leader in circular practices—reducing its carbon footprint, conserving resources, and contributing to a more resilient and sustainable future.

Dr Rahul Mishra
Director KNHPI

Preface

The paper industry, while central to global economic and educational infrastructure, remains a resource-intensive sector. As environmental regulations tighten and sustainability becomes a global priority, there is an urgent need to rethink how industrial waste, particularly from paper mills, is managed. This book, *Paper Industry Waste Processing: A Circular Economy Approach*, brings together innovative research and practical strategies to support the transition from linear to circular waste management practices within the paper industry.

Each chapter of this book is contributed by experienced researchers and professionals in the field.

The journey begins with an introduction (Chapter 1) setting the context for the challenges and opportunities in waste management within the paper sector.

Chapter 2 explores the potential of paper mill sludge in fiber and filler recovery and its reuse in producing low-grade paperboards—a critical look at resource conservation and value addition.

Chapter 3 further examines the use of paper mill sludge in the production of building materials, demonstrating the cross-sectoral benefits of industrial symbiosis.

The book broadens its scope in Chapter 4, with the details of the transformation of coal fly ash into pigments and fillers suitable for papermaking—an innovative path toward resource efficiency and pollution reduction.

Chapter 5 continues with a contribution on energy recovery and decarbonization strategies, highlighting how paper industry wastes can be rechanneled into clean energy generation, contributing to global climate goals.

The transition to circularity is comprehensively addressed in Chapter 6, providing a systemic overview of circular economy principles tailored for the paper industry.

Finally, Chapter 7 offers a detailed case study grounding the theoretical and technological insights shared in previous chapters in real-world application and outcomes.

This book is intended for professionals, researchers, and students engaged in waste management, environmental engineering, sustainable industrial practices, and policy-making.

We hope it serves as a useful guide and catalyst for change toward a more sustainable and circular paper industry.

Author Biographies

Suresh Sundaramurthy is the Head of the Department of Chemical Engineering & Faculty In-Charge, Central Research Facility at Maulana Azad National Institute of Technology Bhopal (India). He completed his Ph.D. in chemical engineering from IIT Roorkee and has held postdoctoral and visiting faculty and research positions at international and national institutions (CUNY, USA, AIT Thailand, JNCASR, Bengaluru, IIT Kanpur, Pondicherry University). His research interests include catalysts, reactor design, environmental biotechnology, waste-to-energy, nanoengineered materials, biofuels, decarbonation, and green hydrogen. He has supervised numerous Ph.D. and master's students and has substantial experience in R&D projects. He has undertaken several energy and environment consultancy projects and technical services, including Gas Authority of India Limited, NCAP-MoEF&CC, Central Effluent Treatment Plant (Indore), CITIIS 2.0 (Ujjain), M.P. Pollution Control Board, Directorate of Urban Administration & Development, State Co-operative Dairy Federation, Animal & Poultry Development Corporation, and C-STEP Pvt Ltd. He has published over 130 research papers in SCI-indexed journals, authored/edited 14 books and 41 book chapters/encyclopedia entries, and holds four patents. He has been a technical expert member of A.I.C.T.E and Bureau of Indian Standards. He has hosted and led different technical industry-academia conferences/webinars, including CO_2 Reduction Processes for Future Energy Development, involving participation from different industry experts with emphasis on the conversion of CO_2 into value-added products, coal gasification, and H_2 as the source for future energy. He has received awards from SERB-IUSSTF, the Ministry of Education, DST-Uttarakhand Council of Science and Technology, and the Institute of Engineers. He was honored as fellow member of the Society of Energy Engineers and Managers & Institution of Engineers (India).

Saakshy completed her M.Tech. and Ph.D. in chemical engineering from Malaviya National Institute of Technology (MNIT) Jaipur and post-graduation in pulp and paper from the Department of Paper Technology, IIT Roorkee (Saharanpur campus). She is presently working in the research wing of the Kumarappa National Handmade Paper Institute Jaipur and handling the Paper Testing and Environment Laboratory. Her research interests include wastewater treatment, pulp and paper technology, handmade papermaking, application of natural dyes, coating technology, paint technology, design and development of machines, development of standards of paper, and alternative lignocellulosic raw material for papermaking. She has more than 1.5 years experience as quality control in-charge in M/s Rama Paper Mills Ltd. and 25 years of research experience in the pulp and paper sector. She is a technical and alert member of the CHD 15 Paper & Its Products Sectional Committee and CHD 14 Printing Inks, Stationery and Allied Products Sectional Committee of the Bureau of Indian Standards and a life member of the Indian Institute of Chemical Engineers. She is ISO convenor of Working Group 30 for Tensile Properties from India and ISO expert member of Working Group 17 for Kraft Liquor Analyses. She has guided Ph.D., M.Tech., and B.Tech. students and guided one Fulbright Fellow from the United States in their thesis, “Application of Natural Dyes in Handmade Paper”. She has one patent awarded and four patent applications filed. She was honored with various prestigious awards for her research contributions from Shri Narayan Rane, Minister of MSME; the Young Scientist award from Dr. Harsh Vardhan, Minister of Science & Technology; and the Best Innovative Product award in 2017. She has completed various research projects as the principal investigator/co-project investigator sponsored by various government organizations, has authored chapters in various books, and has had articles published in national/international journals.

Anuj Kumar Verma presently serves as the Technical Director and is involved in new product development at Aaradhya Pulp & Minerals in Kanpur. He obtained his B.Tech. in chemical engineering from NIT Warangal, followed by an M.Tech. through an exchange program in pulp and paper engineering from both IIT Roorkee and the Technical University of Darmstadt in Germany. He completed his Ph.D. at MANIT Bhopal. His research interests encompass specialty chemicals for the paper industry, artificial pulp, clay-polymeric coatings, and various mineral fillers utilized in the paper industry.

Arisutha Suresh is the Managing Director, M/s Ecoscience & Technology, Bhopal. Dr. Arisutha Suresh was the recipient of post-doctoral fellow from the Asian Institute of Technology (AIT) Thailand. She has received a Ph.D. in energy engineering from the Maulana Azad-National Institute of Technology (MA-NIT) Bhopal, India. She has received a B.Tech. (electrical engineering), M.Tech. (electrical drives and control), and M.B.A. (human resource management) from Pondicherry University. She also completed a Hindi bachelor’s degree course. Dr. Arisutha has more than 15 years’ experience in teaching/research from Lakshmi Narain College of Technology (LNCT) Bhopal; MA-NIT Bhopal; Christ College of Engineering and Technology, Pondicherry; Krishnaswamy College of Engineering and Technology, Cuddalore; and Electricity Board & Whirlpool Software Technologies, Pondicherry. She was invited for technical skill certificate training on AI and renewable energy. She has more than 25 research journal publications in electrical machines, bioenergy, solar energy, hybrid energy systems, control systems, and human resource management and more than 20 presentations/proceedings, some with co-authors. She has published two international books and more than 18 international book chapters from Elsevier (ScienceDirect), Taylor & Francis (CRC Press), Springer, and Encyclopedia-John Wiley & Sons. She has been invited to more than ten technical talks and has attended more than ten seminars/workshops. She worked in different government and industrial funding projects during her deputation at AIT Thailand, MANIT Bhopal.

Acknowledgments

Suresh Sundaramurthy

I am happy to thank a lot of students and teachers for their insightful help during the writing of this book manuscript. I want to thank my mentors for their invaluable support, including Professor S Sundaramoorthy (Retired), Department of Chemical Engineering, Puducherry Technological University (India); Professor I M Mishra (Retired) & Professor VC Srivastava, Department of Chemical Engineering, Indian Institute of Technology Roorkee (India); Professor Teresa J. Bandosz, the City College of New York (USA); Professor CNR Rao, International Centre for Material Science (India); and Professor PK Bhattacharya (Retired), Department of Chemical Engineering, Indian Institute of Technology Kanpur (India). Their straightforward and enthusiastic instruction in the classroom encouraged me.

I would especially like to thank all my friends, colleagues, and collaborators for constant encouragement over the course of the writing and revising. I would like to thank my parents; my wife, Dr. S. Arisutha; and my kids, Krithi and Nithiyasri, for their unwavering support and perseverance throughout this book.

Saakshy

I would like to express my deepest gratitude to all those who have supported and guided me throughout the journey of creating this book, including my co-editors, Dr Suresh Sundaramurthy and Mr. Anuj Kumar Verma.

I am highly thankful to my mentors Dr A K Ray, IIT Roorkee; Dr Mukesh Bansal, IIT Roorkee; Dr Akhilendra Bhushan Gupta, MNIT Jaipur; Dr Kailash Singh, MNIT Jaipur; Dr Ashish Sharma, Director CPPRI; Dr Bipin Thapliyal, Secretary General IARPMA; Dr Rakesh Kumar Jain, UNIDO expert; and Dr Rahul Mishra, Director KNHPI, for their valuable guidance. I would like to thank my father, Dr Khem Chand; my mother, Mrs Kiran Bala; my brother, Dr Salil Agarwal; and my bhabhi, Mrs Neha Agarwal, who have been a continuous support and inspiration to me during the journey of writing this book. I thank them for their patience, understanding, unwavering encouragement, and constant belief in me during the long hours of writing and revision.

Anuj Kumar Verma

First, I would like to thank my co-editors Dr Suresh Sundaramurthy and Dr Saakshy for helping me in writing this book. I would not have been able to write this book without the support and patience of my wife and kids.

1 Status of the Paper Industry
A Circular Economy Approach

Saakshy

1.1 INTRODUCTION

The word paper is derived from the Greek word "papyrus". Papyrus is a plant, and in ancient times, the bark of the papyrus plant was used to convey written messages from one person to another and one place to another. Handmade paper was the first paper made, and the traditional art/technology of handmade papermaking is still a potential export-oriented sector in India, made by the Kagzi clan. Kagzis used pounding and hand sheet-making methods using vats of various designs and made handmade paper. Apart from India, the traditional art of handmade paper is still protected in various countries, including Philippines, Nepal, and China. India is the major player in the art of handmade papermaking next to China. In India the 1990s, the demand for paper started to grow and making paper by hand was replaced by the paper mills in India. The paper industry became more automated with time, and the speed of paper machines reached up to 1800 meters per minute, with designed speeds up to 2000 meters per minute, yielding up to 4500 tons of paper per day [1], with web width ranging from 5 to 10 meters.

1.2 OVERVIEW OF THE GLOBAL PAPER INDUSTRY

According to IBIS world data, there are approximately 45,532 paper mills globally producing 420 million tons of pulp [2]. There are approximately 900 pulp and paper mills, with 5.0% of the total world production. India is the fastest-growing paper market, with an 8.2% rise and a major role played by packaging paper. The rise in Fast Moving Consumer Goods (FMCG) products, textiles, pharmaceuticals, and healthcare has added to the market of packaging paper [3] globally. According to World Bank data, the GDP growth rate is 7.6% annual change. The consumption of paper and paperboard for different countries can be seen in Figure 1.1.

1.2.1 Per Capita Consumption of Paper

The per capita consumption of any country depends on the literacy rate, market, demand for packaging paper, and overall economic growth. The demand for paper in India is growing at a rate of 6 to 7% per annum. The demand for packaging paper

DOI: 10.1201/9781003461395-1

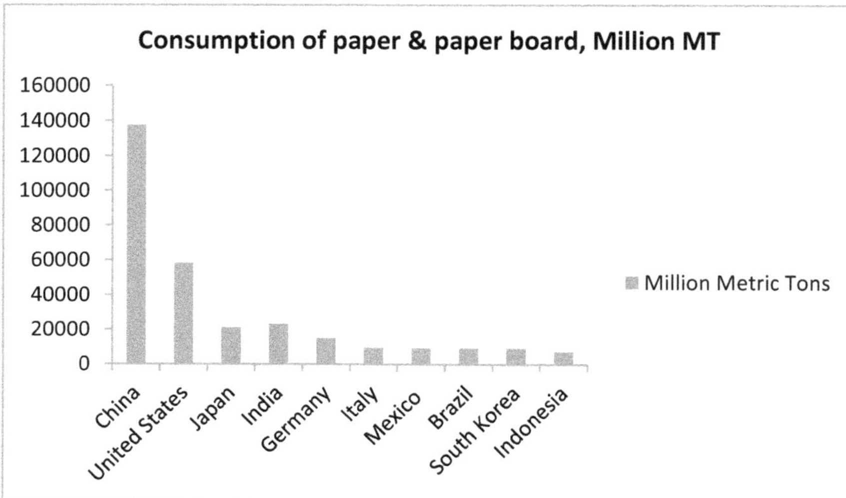

FIGURE 1.1 Consumption of paper and paperboard country wise in 2023 [4].

Source: www.statistica.com

depends on the market for FMCG products. Data on per capita consumption of paper and paperboard from various countries is given in Table 1.1.

In the Asia-Pacific region and Latin America and the Caribbean, production increased by 2% in 2023. Africa remained stable, while Europe and North America recorded declines of 11% and 8%, respectively.

1.2.2 Distribution of Paper Production

In 2023, the regional distribution of paper production is given in Table 1.2.

The distribution of the production of paper in different countries is given in Figure 1.2.

The production of pulp and paperboard worldwide was 401 million tons in 2023. Asia's share of global paper production is 50.6%. Per data from the European Confederation of the Paper Industry (CEPI) [6], global pulp and paper production in 2023 saw a significant decrease compared to the previous year, with a reported decline of around 12.8% in paper and board production, primarily due to adverse economic trends and reduced demand across various paper grades; this translates to a total production of approximately 74.3 million tons in 2023.

With respect to international trade, about one quarter of production is exported. Global trade remained quite stable at around 114 million tons from 2019 to 2022 but dropped by 9% in 2023. The two largest paper and paperboard producers in 2023 were China (134 million tons) and the United States (61 million tons). Their combined production accounted for 48% of global production. The other three

TABLE 1.1
Per Capita Consumption of Paper and Paperboard for Various Countries

Country	Kg in 2018 (Source: FAO 2018) [5]	Kg in 2023 Consumption in Million Metric Tons/Population (Source: statistica 2023; cia.gov)	% Increase (+)/ Reduction (-) in per Capita Consumption of Paper/Paperboard	GDP Growth Rate,%	Literacy Rate,%
Germany	250	177	(-) 29.2% reduction	(-) 0.3	99
South Korea	190	174	(-) 8.42% reduction	1.4	98
Japan	200	172	(-) 14% reduction	1.9	99
USA	229	171	(-) 25.32% reduction	2.5	100
Italy	175	156	(-) 10.85% reduction	0.7	99
China	85	97	(+) 14.11% increase	5.2	99.8
Mexico	74	72	(-) 2.7% reduction	3.2	94.40
Brazil	40	42	(+) 5% increase	2.9	94.69
Indonesia	14	26	(+) 85.71% increase	5.05	99.87
India	13	16	(+) 23.07% increase	8.2	77.7

TABLE 1.2
Regional Distribution of Paper Production

S.N.	Region	Million Tons	%
1	Asia-Pacific	217	54
2	Europe	88	22
3	Northern America	69	17
4	Latin America and the Caribbean	23	6
5	Africa	03	1
	Total	400	

largest producers were Japan (22 million tons), Germany (19 million tons), and India (17 million tons), which accounted for another 15% of global production. Among the top five producers, China was the only country that increased production significantly (20%) over the observed period. Production in India remained about the same from 2019 to 2023, while the United States, Japan, and Germany saw a decline of 11% to 16%. The pulp and paper production of the top five countries is given in Table 1.3.

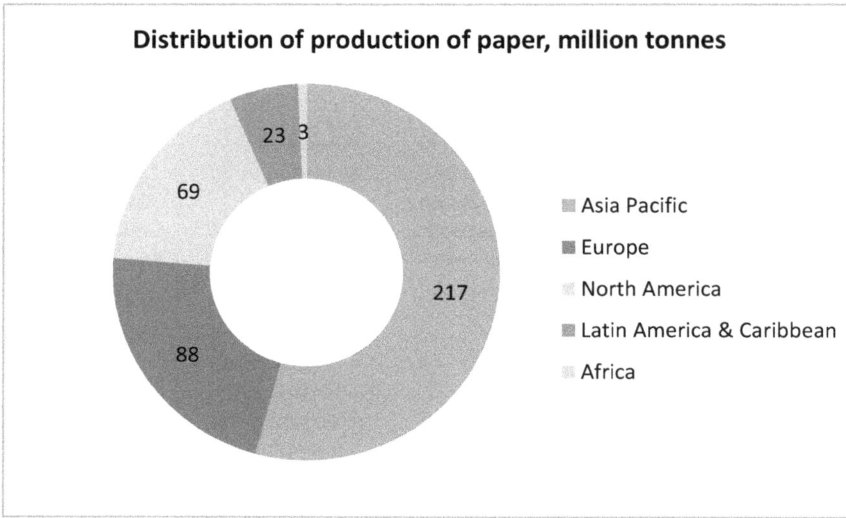

FIGURE 1.2 Distribution of production of paper.

(Source of data: FAO 2023)

TABLE 1.3
Pulp and Paper Production of the Top Five Countries

	China	US	Brazil	Japan	Canada
Wood pulp	26,900	37,364	24,252	7117	12,600
Other fiber	61,330	0		4	0
Dissolving grade pulp		0	1070	163	0
Recovered fiber		28,387	4300	14,920	0
Pulp and paper production, metric tons	88,230	65,751	29,622	22,204	12,600

Source: FAO 2023 (doi.org/10.4060/cd2190t)

1.2.3 Shares of Production and Consumption of Paper and Paperboard

It can be seen in Table 1.4 that the major share of production and consumption of paper and paperboard in 2022 is in Asian countries. The consumption and production of paper and paperboard are given in Table 1.4 and Figure 1.3.

It is obvious from the Figure 1.3 that pulp and paperboard consumption is more than production in Latin America, Asia, and other European country, thus showing the potential for import from other countries.

TABLE 1.4
Consumption and Production of Paper and Paperboard

S.N.	Country	Pulp and Paperboard Production in 2022	Pulp and Paperboard Consumption in 2022
1	Latin America	5.6%	7.3%
2	Asia	49%	50.7%
3	North America	17.8%	17.2%
4	Other European country	4.4%	5.4%
5	CEPI	20%	17.3%
6	Rest of the world	3.3%	2.1%

Source: CEPI [6]

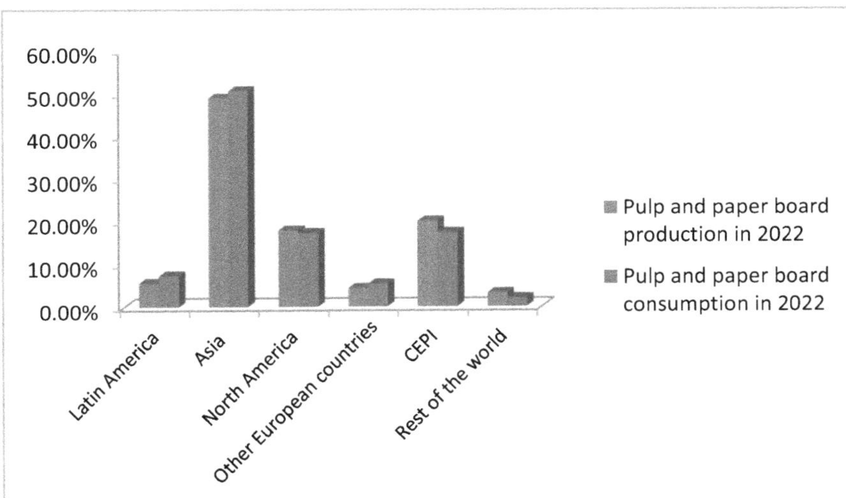

FIGURE 1.3 Consumption and production of paper and paperboard.

1.2.4 Export Data

Total paper and paperboard production in Europe fell by 4.9% in 2022 to 95.5 million tons. High energy costs and raw material prices impacted much of the region. The apparent consumption of paper and paperboard remained unchanged in 2022 at 89.9 million tons. During the second half of 2022, numerous paper-machine closures were driven by a decline in consumption. The latter was the result of higher raw material prices and lower advertising budgets for print products. The production of graphic grades fell by 9.6% to 24.2 million tons in 2022, with uncoated wood-free paper down by 5.9% to 7.4 million tons and coated paper down 10.6% to 8.3 million

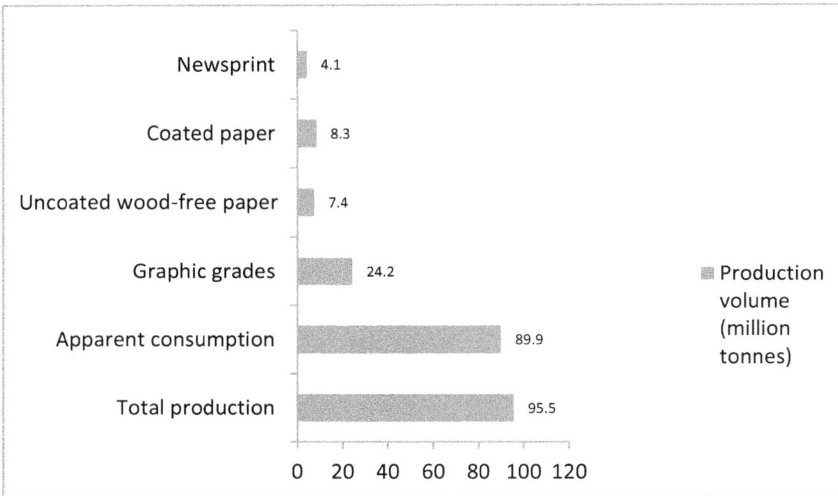

FIGURE 1.4 European paper and paperboard production in 2022.

tons. Newsprint production decreased by 10.0% to 4.1 million tons. The European paper and paperboard production in 2022 is shown in Figure 1.4.

The production of sanitary and household papers rose by 1.1% to 8.1 million tons. The production of packaging materials fell by 2.8% to 55.8 million tons, mainly because of high energy prices. Case material production fell by 3.5% in 2022 to 34.0 million tons, while carton board decreased by 2.0% to 11.7 million tons. Apparent consumption of graphic grades decreased by 0.7% to 22.2 million tons in 2022. Uncoated wood free paper increased by 0.1% to 7.0 million tons, while coated papers decreased by 5.0% to 6.8 million tons. Apparent consumption of newsprint rose by 3.2% to 4.2 million tons. Apparent consumption of sanitary and household papers increased 0.4% in 2022 to 8.1 million tons. Apparent consumption of packaging materials rose by 1.0% to 52.2 million tons in 2022. This positive trend was driven by increased carton board consumption, up 6.7% to 9.4 million tons. Case materials, in contrast, experienced a decline of 1.2% to 33.7 million tons, explained in part by lower levels of online shopping as consumers emerged from COVID-19. Exports of packaging papers fell by 3.8% to 36.1 million tons, while imports rose by 2.7% to 32.5 million tons. Both figures were the result of high energy prices and lower fiber availability from the Russian Federation. The paper production trend in 2022 is shown in Figure 1.5, and the reasons for the increases and decreases in production, consumption, export, and import in 2022 are shown in Figure 1.6.

1.2.5 Pulp Production

Total wood pulp production in 2022 fell by 1.6% to 39.0 million tons, with chemical grades down 1.9%. The production of market pulp was down by 3.6% to 14.4 million

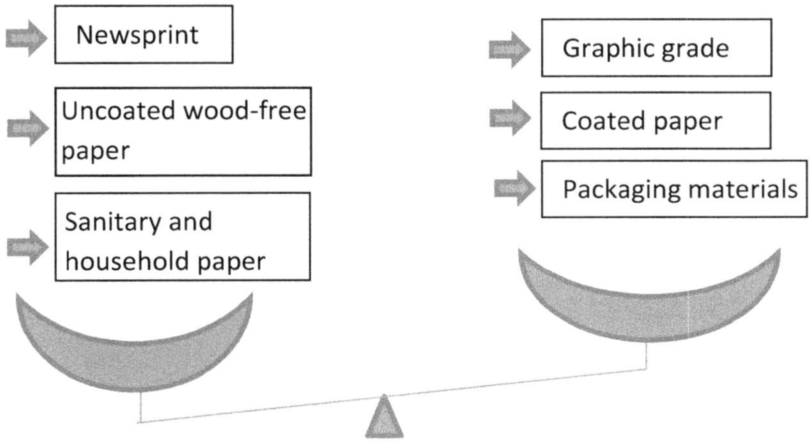

FIGURE 1.5 Paper production trends in 2022.

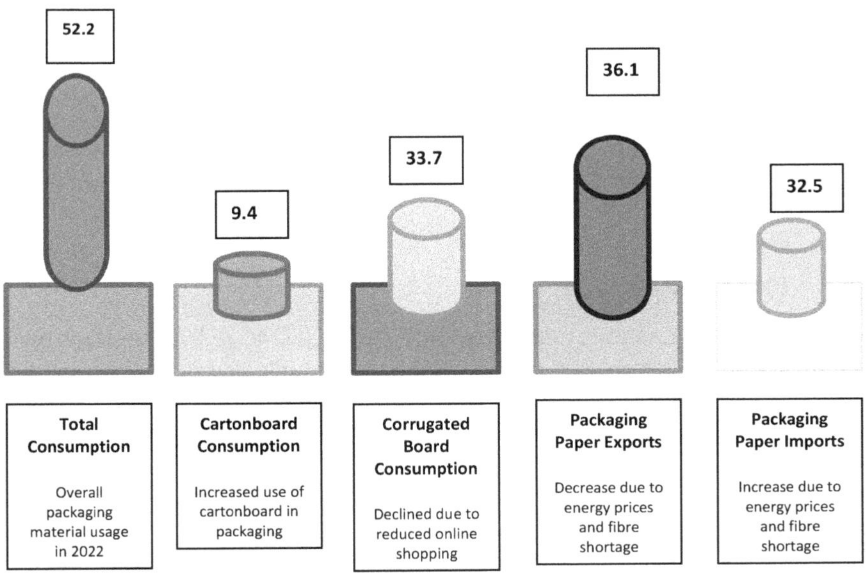

FIGURE 1.6 Reasons for decreases and increases in export, import, and consumption in packaging products in 2022.

tons in 2022 because of high energy prices and lower log availability from the Russian Federation. As a result, wood pulp imports rose by 4.9% to 20.3 million tons in 2022. Wood pulp exports increased by 2.8% to 16.6 million tons in part because of shuttered integrated paper mills finding markets for their pulp production. Paper mills that no longer ran their machines still had fiber-supply contracts for which they

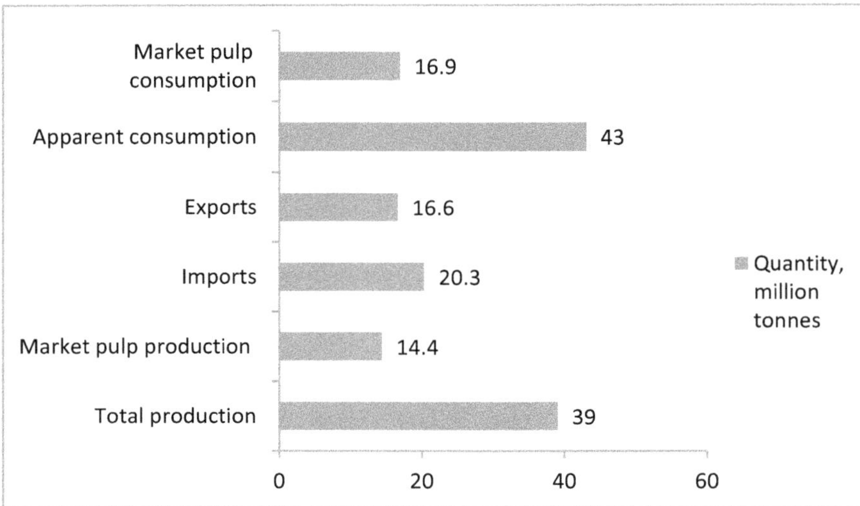

FIGURE 1.7 Wood pulp production, imports, exports, and consumption in 2022.

saw a benefit selling pulp. Apparent consumption of wood pulp fell by 0.1% to 43.0 million tons in 2022, with market pulp increasing by 3.1% to 16.9 million tons. The production and consumption of wood pulp, along with export and import, in 2022 is given in Figure 1.7.

1.2.6 Recycling

The use of paper for recycling decreased by 6.4% in 2022 to 47.5 million tons owing to lower packaging production as well as the record high electricity, gas, and CO_2 prices. The latter disproportionally affected paper recycling mills. Collections fell by 5.0% to 52.6 million tons. Meanwhile, the recycling rate fell to 70.5% in 2022, down from 72.8% in 2021.

1.3 MARKET OUTLOOK

1.3.1 Europe

In Europe, the outlook for the production of paper and paperboard in 2023 showed a decline because of a decline in demand and prices, resulting in numerous machine closures, while wood pulp production was forecast to rise. The apparent consumption of paper and board as well as wood pulp declined owing to reduced economic activity, including destocking throughout the supply chains, as prices for virtually all products declined. The production and consumption forecasts for 2023 in Europe are given in Figure 1.8.

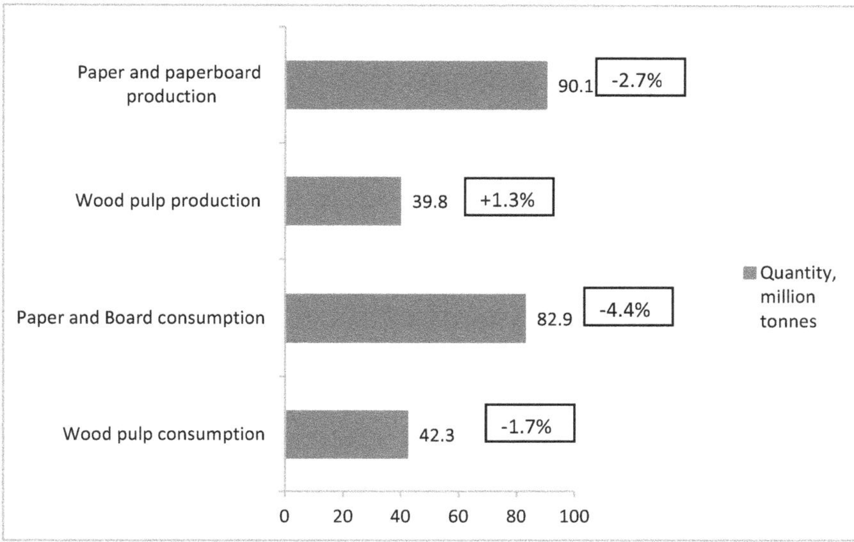

FIGURE 1.8 Production and consumption forecasts for 2023 in Europe.

TABLE 1.5
Paper Production Trends in North America

Parameters	2022	2023
Production volume	Higher	Lower
Market demand	Strong	Weaker
Prices	Stable	Falling

1.3.2 North America

The decline in production of paper and board in North America in initial phase of 2023 in comparison to 2022 was due to less demand and a decline in prices [7] (www.unece.org), which can be seen in Table 1.5.

The decline in various grades of writing- and printing-grade paper in North America is shown in Figure 1.9, with uncoated wood-free paper showing a major decline.

1.4 STATUS OF MARKET OF TOP PAPER COMPANIES

The Indian paper industry accounts for almost 5% of world's paper production. The estimated turnover of the Indian paper industry is over INR 80,000 crore. Approximately 21% of Indian paper mills are based on wood, 71% on recycled fiber, and 8% on agro-residue as raw material. The data for the top ten paper companies in the world is given in Table 1.6 and Figure 1.10.

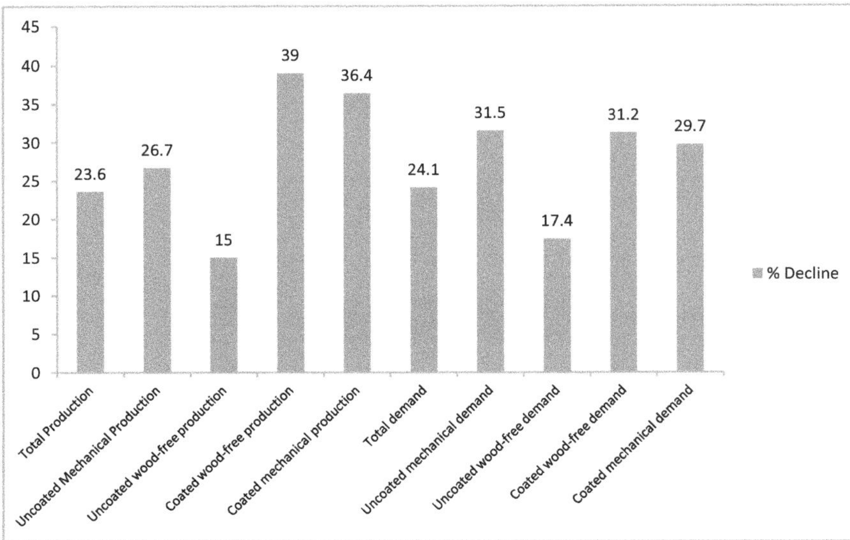

FIGURE 1.9 Decline in North American writing and printing paper industry.

TABLE 1.6
Top Ten Paper Companies in the World (www.statistica.com)

S.N.	Paper Company	Country	Billion U.S Dollars
1	Kimberly Clark Corporation	United States	20.43
2	West Rock	United States	20.30
3	International Paper	United States	18.92
4	Smurfit Kappa Group Plc.	Irish American	12.30
5	UPM-Kymmene Oyj	Finland	11.46
6	Oji Paper Company	Japan	11.01
7	Stora Enso Oyj	Finland	9.4
8	Masco	United States	7.97
9	Nine Dragons Paper Holdings Limited	China	7.82
10	Weyerhaeuser Company	United States	7.7
11	Nippon Paper Group	Japan	7.44
12	Universal Forest Products	United States	7.22
13	West Fraser Timber Co Ltd.	Canada	6.45

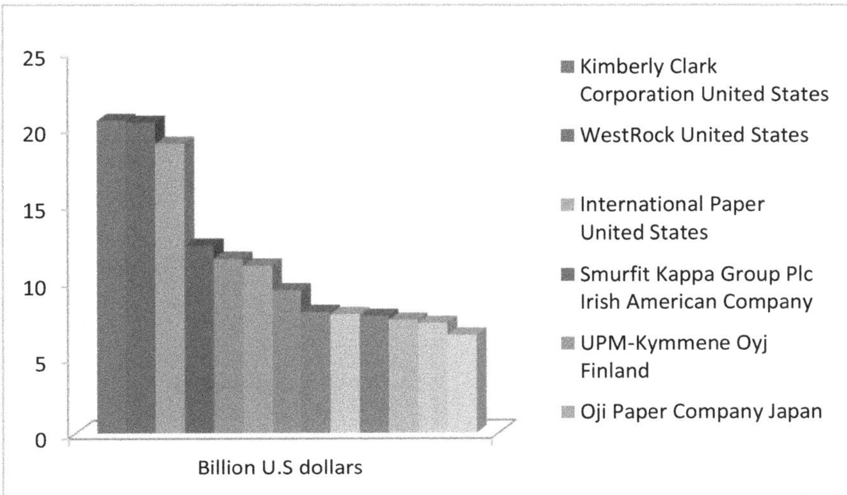

FIGURE 1.10 Market of top paper companies in world.

1.5 RAW MATERIAL USED IN PAPER INDUSTRY

The raw material for the paper industry should have a maximum amount of cellulose content and a minimal amount of lignin content. The raw material for the paper industry is further classified into virgin and recycled fiber. When the source of raw material is waste paper, it is termed "recycled fiber". The paper industry is classified on the basis of raw material: wood based, agro based and recycled fiber based. The raw material for papermaking is mainly divided into woody and non-woody raw material. Woody raw material is further classified into hard and soft woods. Non-woody raw material includes agricultural residues and bast fibers. Bast fibers include jute, ankara, sunn hemp, ramie, flax, etc. and grasses include wheat straw, rice straw, esparto grass, bamboo, etc. There is another grade called recycled or secondary fiber that includes office waste, road sweepings, old newsprint (ONP), and so on. The classification of raw material used in the paper industry is given in Figure 1.11.

1.6 PULPING METHODS

The various pulping methods used in the paper industry are given in Figure 1.12.

1.7 GRADES OF PAPER

There are various grades of paper, which are classified based on their applications. Writing and printing paper, packaging, tissue paper, towel paper, wall paper, insulation paper and board, newspaper, copier paper, art chrome, currency paper,

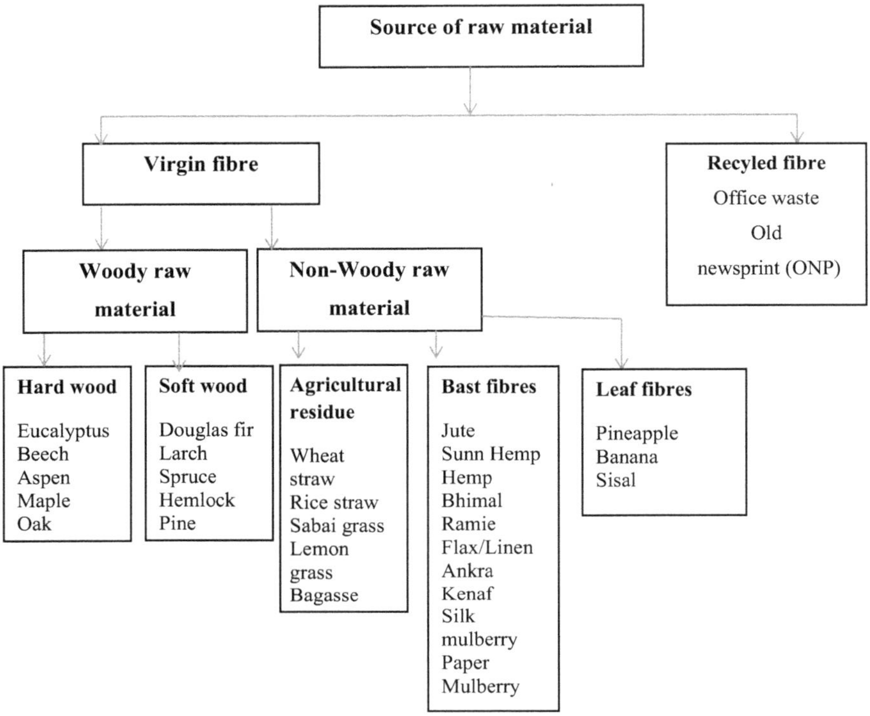

FIGURE 1.11 Classification of raw materials used in the paper industry.

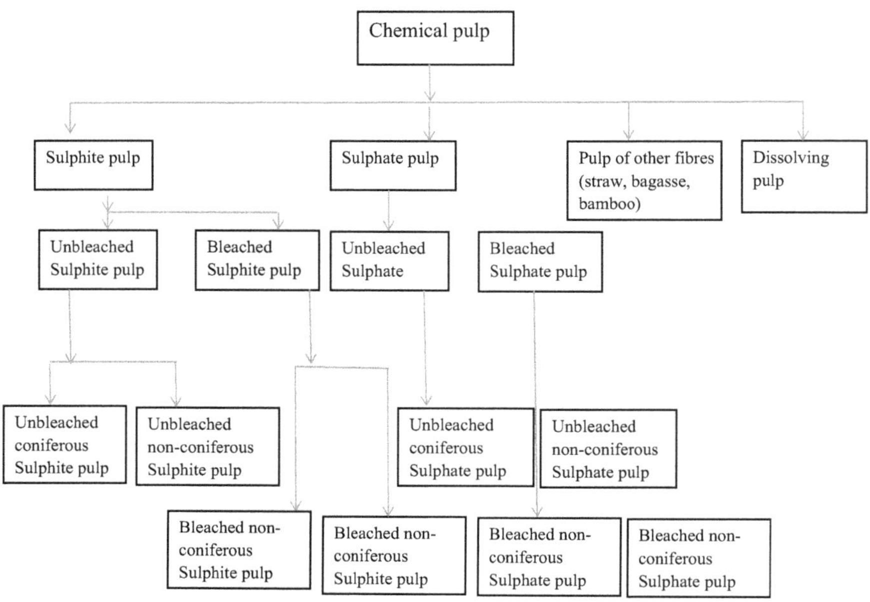

FIGURE 1.12 Classification of chemical pulp.

cheque paper. Toilet paper, facial tissue, emery paper, sand paper, blotting paper, greaseproof paper, glassine paper, vegetable parchment paper. The increase in demand and production of wrapping and packaging grade of paper from 55% share in 2018 to 64% in 2023 and sanitary and household from 8% to 10% while reduction observed in demand for writing and printing grade from 26% share in 2018 to 18% in 2023 [8].

The global pulp and paper market share of production is given in Table 1.7 and Figure 1.13; global production of various grades of paper, in million tons, is given in Figure 1.14.

TABLE 1.7
Global Pulp and Paper Market Share in Production

S.N.	Grade of Paper	% Share	Million Tons
1	Wrapping and packaging	64%	259
2	Writing and printing	18%	73
3	Household and Sanitary	10%	40
4	Other paper and paperboard	5%	20

Source: www.fortunebusinessinsights.com

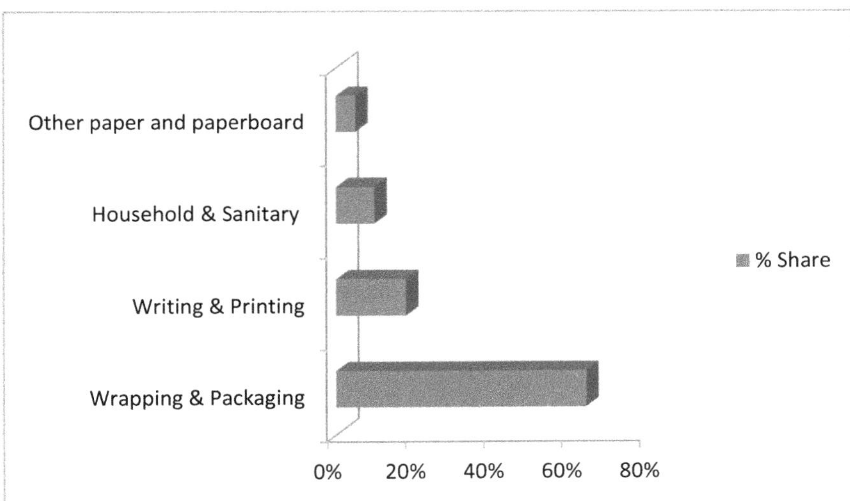

FIGURE 1.13 Global percentage share of production of various grades of paper.

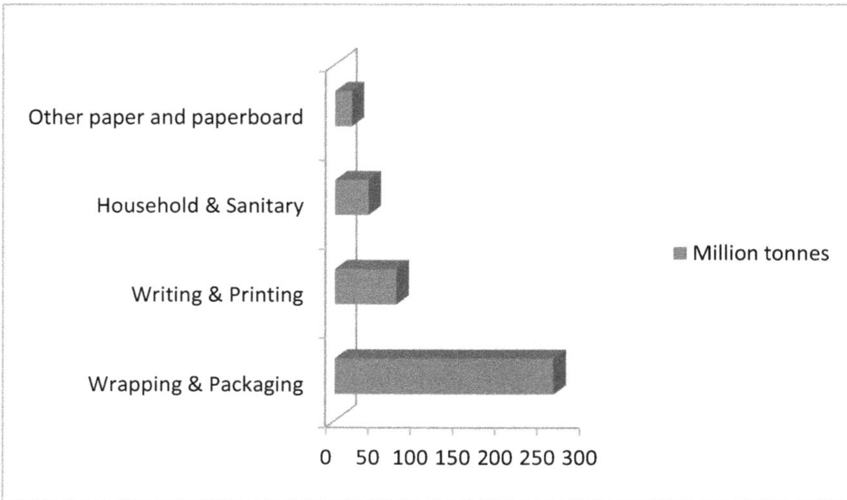

FIGURE 1.14 Global production of various grades of paper in million tons.

1.8 CONCLUSION

Per the data, worldwide, approximately 420 million metric tons of paper are produced per day, making up 13–15% of total wood consumption. The major part of paper production is packaging paper due to the growing market in e-commerce and FMCG products. Writing and printing paper is showing a slight decline every year. The paper industry is the fourth largest when it comes to energy consumption (www.pulpandpaper.com), and continuous efforts are being made in energy reduction and pollution control strategies and policies for the paper industry. The paper industry is sub-categorized on the basis of raw material used into waste paper–based, agricultural residue–based, and wood-based paper industries. The paper industry is also sub-categorized based on production in metric tons per day as small, medium, and large. The paper industry can be seen as a barometer of civilization or the education status of a country. Paper consumption has increased as people have raised their education status or become more civilized.

REFERENCES

[1] www.voith.com
[2] bir.org
[3] www.business-standard.org
[4] www.statistica.com
[5] www.fao.org
[6] CEPI-Confederation of European Paper Industries. (2014). Resource efficiency in the pulp and paper industry: Making more from our natural resources. https://www.cepi.org/resource-efficiency-in-the-pulp-and-paper-industry-making-more-from-our-natural-resources/

[7] Haggith M. Reducing paper consumption and ensuring fair access. In: Martin J, Haggith M, editors. The state of the global paper industry, shifting seas-new challenges and opportunities for forests, people and the climate. Environmental Paper Network, North America, 35.
[8] www.unece.org

2 Paper Mill Sludge Utilization in Fiber, Filler Recovery, and Low-Grade Paperboard Production

Suresh Sundaramurthy and Anuj Kumar Verma

2.1 INTRODUCTION TO PAPER MILL SLUDGE: COMPOSITION AND CHALLENGES

2.1.1 Overview of Sludge Generation in Paper Mills

Paper mill sludge constitutes a heterogeneous mixture of materials generated during various stages of pulp and paper manufacturing processes. Primary sludge originates predominantly from the initial pulping and screening operations, comprising mainly short cellulose fibers, fillers, and coating materials that have been rejected during the papermaking process. Secondary sludge, alternatively known as biological sludge, results from wastewater treatment systems where microorganisms convert dissolved organic matter into cellular biomass. In modern integrated paper mills, a combination of these two fractions—referred to as mixed sludge—represents the predominant waste stream requiring management and disposal. The quantitative aspect of sludge generation presents a significant operational concern for the industry. For every ton of paper produced, approximately 40–50 kg of dry sludge (70% primary sludge & 30% secondary sludge) is generated, with the exact quantity varying according to several factors, including raw material quality, product specifications, process technology, and wastewater treatment efficiency.

High-grade paper and specialty products typically generate larger quantities of sludge compared to standard grades due to more stringent quality requirements. Mills utilizing recycled paper as their primary fiber source encounter particularly high sludge generation rates—often exceeding 100 kg per ton of product owing to the removal of contaminants and degraded fibers present in the recovered paper feedstock.

The compositional profile of paper mill sludge exhibits considerable variability across different mill operations. Primary sludge typically contains 40–70% cellulosic fibers of varying quality and length; 10–30% inorganic materials (primarily calcium carbonate, kaolin clay, and titanium dioxide used as fillers or coating pigments); and smaller amounts of additives, including sizing agents, strength enhancers, and retention aids. Secondary sludge presents a markedly different composition, with 40–60% microbial biomass, residual organic matter, and significantly lower fiber

DOI: 10.1201/9781003461395-2

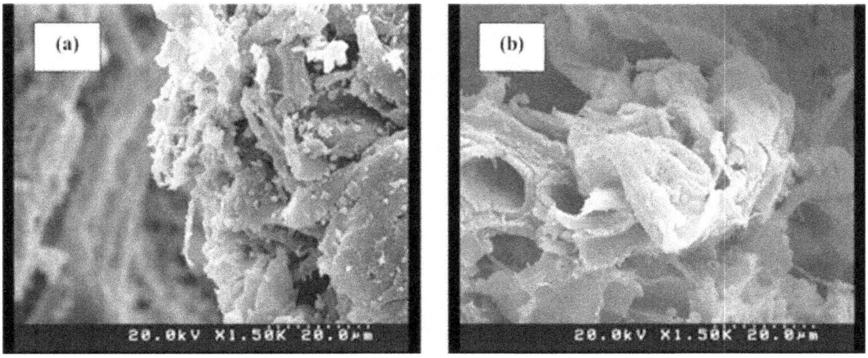

FIGURE 2.1 Cross-sectional microscopy image showing fiber and mineral composition of typical paper mill primary sludge.

content. Both fractions contain substantial water content, with dewatered sludge typically containing 60–80% moisture even after mechanical dewatering operations, presenting considerable challenges for handling and transportation.

2.1.2 Environmental and Economic Implications of Disposal

The traditional management approach to paper mill sludge has centered on disposal methods that increasingly face environmental scrutiny and economic pressure.

Landfilling, historically the predominant disposal route, presents multiple environmental concerns, including potential leachate generation containing organic compounds and trace elements that may contaminate groundwater. The gradual decomposition of organic matter in sludge under anaerobic landfill conditions contributes to methane emissions—a greenhouse gas with global warming potential significantly higher than carbon dioxide. These environmental implications have prompted increasingly stringent regulatory frameworks governing waste disposal across most industrialized nations.

From an economic perspective, conventional disposal methods impose substantial financial burdens on paper manufacturers. Direct costs associated with landfilling have escalated dramatically over the past two decades, with tipping fees in many European countries and North American regions exceeding €100 per ton. This cost trajectory reflects both diminishing landfill capacity and policy instruments designed to discourage landfill disposal, including escalating landfill taxes implemented across numerous jurisdictions. Incineration, while offering volume reduction and potential energy recovery, typically entails even higher operational costs and requires sophisticated emissions control systems to manage potential air pollutants, including particulate matter, acid gases, and trace organic compounds.

The economic equation extends beyond direct disposal costs to encompass additional dimensions including transportation expenses—particularly significant given the high moisture content of sludge—and lost opportunity costs associated with the unutilized material value contained within the sludge. The fiber component represents recoverable cellulosic material that could substitute for virgin or recycled pulp, while mineral fillers constitute potential raw materials for paper production or alternative applications. As virgin fiber and mineral costs continue to rise due to resource competition and supply constraints, the economic incentive for material recovery from sludge streams has strengthened considerably.

Regulatory frameworks governing industrial waste management have evolved significantly, generally following a hierarchical approach that prioritizes reduction, reuse, and recycling over disposal options.

The European Waste Framework Directive and similar legislative structures in other regions have established legal foundations for this hierarchy, with specific targets for industrial waste recovery and restrictions on landfilling of biodegradable waste streams.

Compliance with these evolving regulatory requirements necessitates innovative approaches to sludge management beyond traditional disposal pathways, driving interest in recovery and valorization strategies.

2.2 RECOVERY OF FIBERS AND FILLERS FROM SLUDGE

2.2.1 Mechanical and Chemical Recovery Techniques

The recovery of valuable components from paper mill sludge requires methodologies tailored to the specific characteristics of the material stream and desired end products. Mechanical recovery techniques focus primarily on physical separation processes that exploit differences in particle size, density, or surface properties to isolate fiber and filler fractions. These approaches offer advantages in terms of operational simplicity and avoidance of chemical inputs but may yield lower purity recoveries compared to more sophisticated methodologies.

Screening represents the most fundamental mechanical separation approach, employing vibrating screens with specific aperture sizes to separate fibrous material from smaller filler particles. Industrial applications typically utilize multi-stage screening systems with progressively finer mesh configurations to optimize separation efficiency. Pressure screens incorporating rotating elements that generate hydrodynamic pulses enhance the passage of fine particles while retaining fibrous material. While relatively energy efficient, screening alone rarely achieves sufficient separation quality for high-grade applications due to the intimate association between fibers and mineral particles within the sludge matrix.

Hydrocyclone technology offers more sophisticated mechanical separation based on centrifugal principles. When sludge slurry enters a conical chamber tangentially under pressure, centrifugal forces drive denser particles toward the outer wall, while lighter materials migrate toward the central axis. This differential movement enables separation of heavier mineral components from lighter cellulosic fibers. Industrial systems typically employ hydrocyclones operating in series or cascade arrangements to progressively refine the separation.

The technique achieves moderate separation efficiency but remains sensitive to feed consistency variations and exhibits significant water consumption.

Flotation processes exploit differences in surface hydrophobicity between fibers and mineral components. In dissolved air flotation (DAF) systems, fine air bubbles attach preferentially to hydrophobic fiber surfaces, carrying them to the surface where they form a froth layer that can be mechanically skimmed. Chemical conditioning using surfactants or collectors enhances the natural hydrophobicity differential between organic fibers and inorganic fillers. This approach has demonstrated particular effectiveness for secondary sludge with higher organic content but requires careful chemical management and exhibits sensitivity to slurry pH and conductivity.

Chemical recovery methodologies employ reagents that selectively modify component properties or dissolve specific fractions. Alkaline extraction using sodium hydroxide solubilizes hemicellulose components while leaving cellulosic fibers and inorganic materials largely intact. The resulting fractionation facilitates subsequent separation while potentially improving fiber quality by removing surface contaminants. More aggressive chemical approaches include selective dissolution techniques where specific solvents target either the organic or inorganic fraction. For instance, dilute acid treatments dissolve calcium carbonate fillers while preserving fiber integrity, allowing subsequent separation through conventional mechanical means.

Enzymatic treatments represent an emerging biological approach to sludge fractionation. Specific enzyme combinations including cellulases and hemicellulases selectively modify fiber surfaces, releasing bound mineral particles and facilitating subsequent separation. These biological techniques offer advantages in terms of environmental compatibility and energy efficiency but currently face economic constraints related to enzyme costs and longer processing times compared to conventional chemical approaches.

2.2.2 Efficiency of Fiber and Filler Extraction

The efficiency of recovery processes critically determines both environmental benefits and economic viability of sludge valorization initiatives. Quantitative evaluation typically examines multiple performance indicators, including recovery yield, product purity, energy consumption, and operational stability. These factors collectively determine the technical feasibility and commercial potential of different approaches within specific operational contexts.

Recovery yield—defined as the percentage of target component successfully extracted from the input sludge—varies considerably across different methodologies and sludge types. For fiber recovery, mechanical techniques typically achieve yields ranging from 45–65% of the available cellulosic content in primary sludge, with higher values observed for fresh sludge containing longer fiber fractions. Chemical enhancement through alkaline pretreatment can increase these yields to 60–75% by releasing fibers bound within filler agglomerates. Filler recovery demonstrates different efficiency patterns, with hydrocyclone systems recovering 70–85% of mineral content but at relatively lower purity levels, typically ranging from 80–90%.

The quality of recovered materials fundamentally determines their application potential and economic value. Recovered fibers exhibit shorter length profiles

TABLE 2.1
Comparative Analysis of Fiber Recovery Techniques for Paper Mill Sludge

Recovery Method	Fiber Yield (%)	Energy Consumption (kWh/ton)	Capital Investment	Product Quality
Screening	45–55	150–200	Low	Moderate
Hydrocyclones	50–65	200–250	Medium	Moderate-high
Flotation	55–70	300–400	Medium-high	High
Enzymatic	60–75	100–150	Medium	High
Chemical	65–80	400–500	High	Very high

compared to virgin or conventional recycled pulp, with weighted average lengths typically 0.6–0.8 mm for primary sludge and significantly lower values for secondary sludge sources. Strength properties, particularly tensile and burst indices, generally register 20–30% below those of conventional recycled fiber, necessitating blending with higher-quality fiber sources for many applications. Brightness characteristics depend significantly on the original sludge source, with fibers recovered from deinking sludge presenting particular challenges due to residual ink particles resistant to separation.

Energy intensity represents a critical factor in determining process sustainability and operating economics. Mechanical separation techniques generally exhibit lower energy demands compared to chemical recovery approaches that require heating and agitation during reaction phases. Process water consumption and chemical inputs contribute significantly to both environmental footprint and operational costs. Advanced systems incorporating water recirculation loops and chemical recovery systems demonstrate substantially improved resource efficiency but require higher capital investment and operational complexity.

Scale considerations significantly impact recovery efficiency and economic viability. Laboratory-scale demonstrations frequently achieve higher separation performance than industrial implementations due to more precise control of operating parameters and shorter residence times that minimize reaggregation phenomena. Successful industrial applications require robust process design accommodating variability in sludge characteristics and effective integration with existing mill operations to leverage available infrastructure and energy streams.

2.2.3 Case Studies and Pilot-Scale Implementations

Theoretical understanding of recovery processes becomes meaningful through practical implementation at industrially relevant scales. Examination of representative case studies illustrates both the potential and challenges associated with fiber and filler recovery from paper mill sludge under real-world conditions.

The Stora Enso Langerbrugge Mill in Belgium implemented an integrated fiber recovery system processing 15,000 tons of primary sludge annually from their deinking operations. The system combines screening and flotation technologies in a multi-stage configuration, recovering approximately 60% of the fiber content for reintegration into their newsprint production. Economic analysis demonstrated a payback period of 2.4 years based primarily on reduced fiber purchasing costs and decreased waste disposal expenses. Process challenges included managing variability in the incoming sludge stream and addressing buildup of contaminants in the recirculated process water, necessitating installation of additional water treatment capacity.

A contrasting approach was demonstrated at SCA's Witzenhausen packaging mill in Germany, where emphasis was placed on recovering both fiber and mineral components from mixed sludge. Their process employs selective dissolution of calcium carbonate using carbon dioxide injection to form soluble calcium bicarbonate, followed by mechanical separation of the released fibers. The dissolved calcium compound is subsequently precipitated under controlled conditions to form high-purity calcium carbonate suitable for reintroduction as paper filler. This integrated approach achieved fiber recovery rates exceeding 70% while producing mineral products with brightness values above 90% ISO, suitable for direct reintegration into their production system.

A pilot-scale implementation at Norske Skog's Boyer mill in Australia focused specifically on secondary sludge valorization, a traditionally more challenging fraction due to lower fiber content and higher biological component. Their approach combined enzymatic pretreatment using cellulase complexes with subsequent flotation separation. While achieving lower overall recovery rates of approximately 40% compared to primary sludge systems, the process successfully extracted fibers suitable for corrugating medium production. The residual biological fraction underwent anaerobic digestion, generating biogas that partially powered the recovery process, creating an elegant closed-loop system that maximized resource utilization.

Common challenges encountered across these implementations included managing the variability inherent in sludge streams, particularly in mills processing variable furnishes or producing multiple paper grades. Successful operations incorporated robust monitoring systems tracking key parameters including fiber morphology, ash content, and biological activity to enable real-time process adjustments. Integration with existing mill water systems proved critical for overall efficiency, with installations that effectively leveraged existing heat recovery and water clarification infrastructure demonstrating superior economic performance.

The economic viability of these implementations varied considerably depending on local factors, including energy costs, fiber pricing, and waste disposal fees. Operations in regions with stringent landfill regulations and high disposal costs achieved substantially faster return on investment, typically ranging from 1.5 to 3 years. Mills producing higher-value products demonstrated greater sensitivity to recovered material quality, often implementing additional refining stages to enhance fiber properties or mineral brightness characteristics.

2.3 UTILIZATION IN LOW-GRADE PAPERBOARD PRODUCTION

2.3.1 Process Integration for Sludge-Based Paperboard

The integration of recovered sludge components into paperboard manufacturing processes represents a pragmatic pathway for material valorization, particularly for mills producing packaging grades where aesthetic requirements are less stringent compared to graphic papers. Successful integration requires careful consideration of material characteristics, process modifications, and product quality specifications to achieve both technical performance and economic viability.

Process integration strategies typically follow one of three fundamental approaches: direct incorporation of dewatered sludge, utilization of recovered fiber and filler fractions as discrete raw materials, or a combination approach where partially separated sludge undergoes limited processing before introduction into the furnish. The direct incorporation method offers operational simplicity but provides limited control over final product properties due to the heterogeneous nature of the sludge. Conversely, utilizing fully separated components enables precise furnish management but entails higher processing costs. The hybrid approach, where sludge undergoes coarse fractionation without complete component separation, often represents an optimal compromise between process complexity and product control.

Stock preparation systems require specific modifications to accommodate sludge-derived materials. Recovered fibers typically exhibit higher fines content and reduced drainage properties compared to conventional pulp, necessitating adjustments to refining parameters.

Lower refining intensity with extended residence time generally yields optimal results when processing these materials, avoiding excessive fiber cutting while still developing adequate bonding potential. Screening systems often require reconfiguration with coarser apertures to accommodate the different morphological profile of recovered fibers, balancing contaminant removal against yield optimization.

Wet-end chemistry presents particular challenges when incorporating sludge components due to their variable surface chemistry and higher concentration of dissolved and colloidal substances. Retention aid systems typically require enhancement through combination approaches utilizing both conventional polyacrylamide-based flocculants and microparticle systems to manage the complex interactions between recovered fibers, fillers, and process water components. pH management becomes particularly critical when incorporating chemically treated recovery streams, with buffering systems often necessary to maintain stable operating conditions.

Dewatering behavior on the paper machine typically exhibits higher sensitivity to process variations when utilizing sludge-derived materials. The higher fines content generally reduces drainage rates, potentially limiting machine speed or requiring vacuum system upgrades.

Press section modifications, including extended nip configurations or increased linear loads, can partially mitigate these limitations by enhancing mechanical dewatering before the energy-intensive drying section. Felt conditioning requires particular attention due to the propensity of recovered materials to cause accelerated fabric contamination, with enhanced chemical cleaning regimes typically necessary to maintain optimal performance.

Sheet formation characteristics present both challenges and opportunities when incorporating sludge-derived materials. The shorter fiber profile typically contributes to improved formation uniformity but may simultaneously reduce in-plane strength properties. Modern gap-former configurations demonstrate particular suitability for these furnishes, providing enhanced dewatering capacity while maintaining formation control through precisely managed drainage pulses. Shake system modifications can provide additional formation improvement through optimized drainage profiles tailored to the specific characteristics of recovered fiber components.

2.3.2 Properties and Performance of Boards Using Recovered Materials

The physical, mechanical, and functional properties of paperboard incorporating sludge-derived components demonstrate distinctive characteristics that influence both manufacturing parameters and end-use performance. Understanding these property modifications enables appropriate product positioning and application targeting to leverage the unique attributes of these materials.

Structural properties exhibit systematic variations correlated with sludge incorporation rates. Bulk density typically increases by 5–15% at incorporation rates of 20–30%, reflecting the higher fines content and modified pore structure within the sheet. This density increase translates to reduced caliper at equivalent basis weight, potentially necessitating grammage adjustments to maintain package rigidity specifications. Internal bond strength, measured by z-directional tensile or Scott bond tests, generally shows enhancement due to the increased bonding surface area provided by shorter fibers and fine material, with improvements of 10–25% commonly observed at moderate inclusion rates.

Mechanical performance characteristics demonstrate more complex response patterns. Tensile strength typically decreases gradually with increasing sludge content, showing reductions of 10–20% at 30% incorporation rates due to reduced effective fiber length. This impact can be partially mitigated through refining strategy modifications or strength additive adjustments. Remarkably, compression strength properties including short-span compression test (SCT) and ring crush test (RCT) values often show minimal degradation or even modest improvements at moderate inclusion levels, attributed to the enhanced sheet density and bonding characteristics. This behavior makes sludge-derived materials particularly suitable for packaging applications where compression resistance drives performance requirements.

Surface properties reflect the influence of the high fines and potential filler content in recovered materials. Smoothness typically improves with increasing sludge incorporation, benefiting printing applications requiring uniform ink transfer. However, surface strength may decrease, potentially causing operational issues in converting operations involving high-tack inks or aggressive adhesives.

Water absorption characteristics, critical for conversion processes including corrugating and folding carton manufacturing, typically show faster initial absorption but lower total capacity—a complex profile requiring process-specific evaluation and adjustment.

Functional performance in end-use applications reveals both advantages and limitations of sludge-derived boards. Package rigidity, a critical parameter for stacking strength and consumer perception, typically shows moderate reductions at equivalent basis weight, requiring either grammage increases or structural design modifications to maintain performance specifications. Convertibility indicators including folding endurance and scoring quality demonstrate mixed results, with performance heavily dependent on specific sludge characteristics and preprocessing methods. Barrier

TABLE 2.2
Mechanical Properties of Paperboard with Varying Sludge Incorporation Rates

Property	Reference Board	10% Sludge	20% Sludge	30% Sludge
Tensile index (Nm/g)	45.5	43.2	39.8	36.1
SCT index (Nm/g)	23.1	23.4	22.9	21.8
Internal bond (J/m^2)	195	210	228	234
Burst index ($kPa{\cdot}m^2/g$)	2.85	2.73	2.58	2.38
Bendtsen roughness (ml/min)	320	295	270	245

properties generally degrade with increasing sludge content due to the more heterogeneous sheet structure, potentially requiring additional coating or lamination for moisture-sensitive applications.

Environmental exposure response presents distinctive patterns for sludge-containing boards. Moisture sensitivity typically increases, with dimensional stability measurements showing 10–20% greater hygroexpansivity under cyclic humidity conditions.

Accelerated aging studies indicate generally comparable performance to conventional boards, although products incorporating secondary sludge with higher biological content may demonstrate slightly faster brightness reversion under extended ultraviolet exposure.

2.3.3 Cost-Benefit Analysis

The economic proposition of utilizing sludge-derived materials in paperboard manufacturing encompasses multiple value dimensions including direct material cost savings, disposal cost avoidance, and potential market positioning advantages. Comprehensive assessment requires systematic evaluation of these factors against implementation costs and potential operational impacts.

Direct material cost impacts constitute the most immediately quantifiable benefit. Replacement of virgin fiber with recovered sludge components typically yields raw material cost reductions ranging from €30–50 per ton of finished product at 20% incorporation rates, with specific values dependent on virgin fiber pricing and sludge recovery efficiency. Mineral component recovery contributes additional savings through reduced filler purchasing requirements, although this benefit applies primarily to filled board grades rather than conventional packaging products. Combined material savings generally represent 3–8% of total manufacturing cost, depending on specific grade structure and local fiber economics.

Waste disposal cost avoidance represents a second major economic driver, particularly in regions with progressive waste management regulations. Landfill diversion of sludge translates to direct savings on tipping fees ranging from €40–120 per ton depending on regional regulations and available disposal infrastructure. Additional regulatory benefits may include avoidance of landfill taxes, which have demonstrated consistent upward trajectories across most industrialized nations as policy instruments to discourage disposal. These combined avoidance costs can contribute an additional €20–40 per ton of finished product at typical incorporation rates, substantially enhancing the overall economic proposition.

Implementation costs for sludge utilization systems vary considerably based on technical approach and existing infrastructure. Capital investment for basic systems enabling direct incorporation without extensive separation typically ranges from €1–2 million for a medium-sized mill producing 200,000 tons annually.

More sophisticated recovery systems implementing advanced fractionation technology may require investments of €3–5 million but generally yield higher-quality recovered materials suitable for wider application ranges. Operating costs including energy, chemical inputs, and maintenance typically add €15–25 per ton of processed sludge, partially offsetting the material and disposal savings.

Product positioning considerations introduce more complex economic factors that resist simple quantification. Boards incorporating recovered materials may qualify for environmental certification programs, including recycled content designations or reduced carbon footprint classifications, potentially commanding price premiums in environmentally conscious market segments. Conversely, potential limitations in certain performance characteristics may necessitate market repositioning or application restrictions that impact realizable value. These factors demonstrate high specificity to individual market conditions and customer requirements, necessitating customized assessment for each implementation case.

Return on investment calculations integrating these diverse factors typically yield payback periods ranging from 1.5–4 years for basic implementation systems. Operations in regions with higher disposal costs and stringent regulatory frameworks generally achieve faster returns, while mills with lower waste management costs may require additional value drivers to justify implementation. Scale effects demonstrate significant impact on economic viability, with larger operations typically achieving more favorable economics due to better utilization of capital equipment and enhanced ability to manage product streams across multiple grades and customers.

Risk factors requiring consideration in economic assessment include potential variability in sludge quality affecting product consistency, regulatory uncertainty regarding future waste classification and management requirements, and market acceptance considerations for products incorporating recovered materials. Sensitivity analysis addressing these factors typically indicates that disposal cost trends represent the most significant variable influencing long-term economic sustainability of sludge utilization pathways.

2.4 CIRCULAR ECONOMY AND SUSTAINABILITY PERSPECTIVES

2.4.1 Sludge Valorization in the Context of the Circular Economy

Paper mill sludge valorization exemplifies fundamental circular economy principles through the transformation of production residuals into valuable inputs, closing material loops that would otherwise remain open in conventional linear production models.

This paradigm shift extends beyond simple waste reduction to encompass systemic redesign of material flows, value creation mechanisms, and industrial relationships within the broader paper industry ecosystem.

The conceptual framework of the circular economy applied to sludge management necessitates examination across multiple nested cycles. The primary technical cycle focuses on direct material recirculation within the originating production system, where recovered fibers and minerals directly replace virgin inputs in the same manufacturing process from which they originated. This represents the most efficient implementation of circularity principles, minimizing transportation requirements and ensuring compatibility between recovered materials and production systems. However, this approach faces limitations related to potential accumulation of contaminants or gradual quality degradation through successive recycling cycles.

Secondary technical cycles encompass valorization pathways where sludge components find application in alternative paper industry products, particularly in cases where the originating mill produces grades unsuitable for direct reincorporation of recovered materials. The cascading approach, where materials transition from higher-value to progressively lower-value applications through successive use cycles, provides extended material utility while acknowledging the inherent quality limitations of repeatedly recycled components. This manifestation of circularity principles requires coordinated material flows between different production facilities, potentially facilitated through industrial symbiosis networks or formal exchange platforms.

Biological cycles represent complementary pathways particularly relevant for secondary sludge with higher organic content and limited fiber recovery potential. Anaerobic digestion systems generate biogas that can offset fossil energy consumption within mill operations while producing digestate applicable as soil amendment, effectively connecting industrial and agricultural systems through nutrient cycling pathways. Composting applications similarly return organic matter and minerals to soil systems, completing natural nutrient cycles through integration of industrial residuals into agricultural production systems.

Circular business models evolving around sludge valorization demonstrate innovative approaches to value creation and distribution. Service-oriented models have emerged where specialized technology providers implement and operate recovery systems within mill premises under performance contracts, sharing the economic benefits of recovered materials and avoided disposal costs. This approach alleviates capital constraints while distributing risk between technology providers and mill operators. Regional collaborative models where multiple smaller mills collectively supply sludge to centralized recovery facilities achieve economies of scale unattainable for individual operations, enabling implementation of more sophisticated technologies with enhanced recovery efficiency.

Policy frameworks increasingly recognize and incentivize circular approaches to industrial residuals management. Extended producer responsibility systems incorporating eco-modulation of fees based on recyclability and recycled content provide economic signals favoring closed-loop systems. Green public procurement requirements specifying minimum recycled content create market pull for products incorporating recovered materials, supporting price premiums that enhance economic viability of recovery systems.

These policy instruments collectively strengthen the business case for circular approaches to sludge management beyond direct operational economics.

2.4.2 Life Cycle Assessment and Carbon Footprint Considerations

Quantitative sustainability assessment through life cycle methodologies provides critical perspective on the environmental implications of sludge valorization pathways compared to conventional disposal routes.

This systematic approach encompasses multiple impact categories across the entire value chain, enabling identification of potential burden-shifting between environmental compartments and highlighting leverage points for system optimization.

Carbon footprint analysis typically demonstrates significant advantages for material recovery pathways compared to landfill disposal, primarily due to avoidance of methane emissions from anaerobic decomposition of organic content. Life cycle studies indicate typical greenhouse gas reductions of 0.8–1.2 tons CO_2-equivalent per ton of dry sludge diverted from landfill to material recovery applications. Energy recovery through incineration presents intermediate carbon performance, with emissions avoidance from displaced fossil fuels partially offset by direct CO_2 release from biomass oxidation. The specific carbon intensity of local electricity grids and displaced heating fuels significantly influences comparative performance between material and energy recovery pathways.

Broader life cycle impact assessment reveals more complex sustainability profiles across different environmental dimensions. Material recovery pathways generally demonstrate advantages in fossil resource depletion through virgin fiber substitution and reduced mineral extraction for filler production. Water consumption impacts show mixed results, with recovery processes themselves requiring significant water inputs but simultaneously avoiding the substantial water footprint associated with virgin fiber production. Eutrophication potential typically favors recovery pathways due to reduced landfill leachate generation, although process water management in recovery systems requires careful design to prevent nutrient releases to water bodies.

Land use considerations present multifaceted implications across different time horizons. Immediate impacts favor recovery pathways through avoidance of landfill space consumption, preserving land for alternative uses including carbon sequestration in forest systems. Longer-term perspectives must consider potential soil contamination risks from land application of inadequately treated organic residuals, particularly regarding persistent compounds including certain per- and polyfluoroalkyl substances (PFAS) increasingly identified in paper mill residuals from specific production pathways.

Human toxicity and ecotoxicity assessments present particular methodological challenges due to the complex and variable composition of paper mill sludge.

The heterogeneous nature of potential contaminants requires case-specific analysis rather than generalized conclusions. Recovery processes incorporating thermal treatment stages typically demonstrate reduced toxicity concerns through destruction of organic contaminants, while purely mechanical approaches may preserve these compounds within recovered materials, necessitating careful application selection and potential restrictions for food-contact packaging materials.

Methodological considerations significantly influence life cycle assessment outcomes for sludge management systems. System boundary definition, particularly regarding avoided virgin material production and disposal impacts, critically affects comparative results. Allocation approaches for multi-output systems where recovery processes generate multiple valuable streams require careful justification and sensitivity analysis. Temporal considerations, including carbon accounting methodologies for biogenic carbon flows and long-term landfill emissions, introduce additional complexity, requiring transparent handling within assessment frameworks.

2.4.3 Regulatory Framework and Incentives

The regulatory landscape governing paper mill sludge management has evolved considerably, transitioning from pollution control paradigms focused on limiting environmental harm toward resource efficiency frameworks actively promoting valorization pathways. This evolution creates both compliance obligations and strategic opportunities for paper manufacturers implementing circular approaches to residuals management.

Waste classification frameworks fundamentally determine available management options and associated regulatory requirements. The European Waste Catalogue designation of paper mill sludge as non-hazardous waste (code 03 03 11) enables various recovery pathways without triggering stringent hazardous waste provisions. However, country-specific interpretations regarding the precise boundary between waste and product status for recovered materials create regulatory uncertainty affecting cross-border material flows and market development. The "end-of-waste" criteria concept introduced in advanced regulatory systems provides potential pathways for recovered materials to exit the waste regulatory regime when meeting specific quality standards and intended for defined applications.

Landfill directives and associated economic instruments represent powerful drivers toward recovery pathways. The EU Landfill Directive's progressive restrictions on biodegradable waste disposal, implemented through diminishing allocation allowances, effectively mandates alternative management approaches for high-organic-content sludge fractions. Landfill taxes implemented across most industrialized nations have demonstrated consistent upward trajectories, with current rates in leading European countries exceeding €100 per ton, fundamentally altering the economic equation between disposal and recovery options.

Product regulatory frameworks increasingly incorporate circular economy principles through recycled content requirements and material identification provisions. The European Paper Packaging Essential Requirements specify recyclability criteria while allowing adjusted heavy metal limits for materials incorporating recovered components, acknowledging the technical challenges associated with trace element management in circular systems. Food contact regulations present particular compliance challenges for recovered materials due to uncertainty regarding potential contaminant profiles, with current frameworks generally requiring case-specific migration testing and authorization rather than generic approval pathways.

Financial incentive structures beyond direct regulatory requirements increasingly support sludge valorization initiatives. Green public procurement specifications incorporating minimum recycled content requirements create market demand for products utilizing recovered materials, while sustainability certification systems including ecolabels provide market differentiation mechanisms potentially supporting price premiums.

Climate finance mechanisms including carbon credit systems for landfill methane avoidance provide additional revenue streams for projects demonstrating quantified emissions reductions through diversion from conventional disposal pathways. Research and development support programs at national and supranational levels have evolved to specifically target circular economy implementation challenges.

The European Union's Horizon Europe program includes dedicated funding streams for industrial symbiosis projects and technologies enabling valorization of industrial side streams, while national innovation agencies increasingly prioritize technologies addressing resource efficiency imperatives. These programs provide both financial support for technology development and valuable validation for emerging approaches, reducing perceived implementation risk for commercial adoption.

2.5 FUTURE TRENDS AND TECHNOLOGICAL INNOVATIONS

2.5.1 Emerging Technologies for Enhanced Recovery

The technological landscape for paper mill sludge valorization continues to evolve rapidly, with emerging approaches promising enhanced recovery efficiency, expanded application potential, and improved economic performance. These innovations span multiple technological domains, including separation science, biological processing, and thermal conversion pathways, collectively expanding the toolkit available to paper manufacturers seeking optimal value extraction from sludge streams.

Advanced separation technologies introducing more selective fractionation capabilities represent a prominent development direction. Microfiltration and ultrafiltration membrane systems enable precise separation based on particle size, achieving higher-purity recovered fractions compared to conventional mechanical approaches. These technologies demonstrate particular promise for deinking sludge applications, where removal of submicron ink particles has traditionally limited fiber recovery quality. Industrial implementations remain limited by membrane fouling challenges in the complex sludge environment, though recent developments in antifouling materials and cleaning protocols have demonstrated substantially improved operational stability.

Supercritical fluid extraction using carbon dioxide under high pressure and moderate temperature conditions enables highly selective separation of organic contaminants, including residual adhesives, waxes, and polymerization products that compromise conventional recovery methods.

The process preserves fiber morphology while enhancing brightness characteristics, though current economic viability remains constrained by high-pressure equipment requirements and energy intensity. Research pathways focusing on co-solvent approaches using ethanol-CO_2 mixtures have demonstrated promising results with reduced pressure requirements, potentially addressing the primary economic limitation.

Biological processing innovations increasingly leverage advances in industrial biotechnology to enhance value extraction from sludge components. Engineered enzyme consortia tailored to the specific composition of mill-specific sludge streams demonstrate enhanced effectiveness compared to generic commercial preparations, with optimized combinations of cellulases, hemicellulases, and auxiliary enzymes achieving fiber release while minimizing fiber damage.

Fungal pretreatment systems utilizing white-rot species, including *Trametes versicolor* and *Pleurotus ostreatus*, selectively degrade lignin components and contaminants while preserving cellulosic fibers, though commercial implementation faces challenges related to process time requirements and contamination control in non-sterile industrial environments.

Thermal valorization pathways extend beyond conventional incineration to encompass emerging technologies offering enhanced resource recovery potential. Hydrothermal carbonization (HTC) converts sludge organic content into hydrochar through treatment in subcritical water (180–250°C) under autogenous pressure, producing a coal-like solid with enhanced energy density and potentially valuable adsorption properties. The process effectively handles high-moisture sludge without energy-intensive dewatering, while the liquid phase captures solubilized minerals potentially recoverable as secondary products. Pilot implementations demonstrate particular suitability for secondary sludge with limited fiber recovery potential, providing alternative valorization pathways complementary to material recovery approaches for primary sludge fractions.

Pyrolysis systems operating under oxygen-limited conditions at temperatures of 400–600°C generate three product streams with distinct valorization potential: bio-oil suitable for energy applications or chemical extraction, syngas for process heating, and biochar offering soil amendment properties and carbon sequestration benefits.

Fractional condensation systems enable selective recovery of specific chemical compounds from the volatile stream, potentially accessing higher-value markets compared to pure energy recovery. These thermochemical platforms demonstrate particular promise for managing residual fractions remaining after fiber and mineral recovery, addressing the limitations of purely mechanical approaches while avoiding complete mineralization of the organic content.

2.5.2 Hybrid Approaches Combining Bio, Thermal, and Mechanical Methods

The recognition that no single technology delivers optimal recovery across all sludge fractions and components has driven growing interest in integrated hybrid approaches combining complementary processes. These multi-technology systems enable more comprehensive value extraction through sequential targeting of different sludge constituents, potentially achieving higher overall recovery rates and accessing multiple product markets simultaneously.

Sequential extraction approaches typically begin with mechanical or physico-chemical processes targeting fiber recovery, followed by secondary stages addressing the residual stream.

This configuration preserves the relatively higher-value fiber component in its most intact form while still extracting value from remaining materials. Industrial implementations have demonstrated fiber recovery through screening and flotation followed by enzymatic treatment of the residual stream to generate fermentable sugars for bioethanol production. Alternative configurations employ mechanical fiber recovery followed by anaerobic digestion of residuals, generating biogas while simultaneously reducing the volume of final residuals requiring disposal.

Mineral extraction integrated with organic valorization represents another promising hybrid approach, particularly relevant for sludge streams with significant calcium carbonate content. Selective dissolution of carbonates using either acidification or carbonation (CO_2 injection) releases bound fibers while simultaneously generating a mineral stream. After fiber separation, controlled precipitation regenerates calcium carbonate with tailored properties suitable for paper filling applications or alternative markets including plastics and construction materials. This approach effectively separates the organic and mineral fractions while maintaining the value of both streams, though implementation complexity and sensitivity to sludge composition variability present ongoing challenges.

Thermal-mechanical integration approaches leverage thermal modification to enhance subsequent mechanical separation efficiency. Steam explosion pretreatment, where sludge undergoes brief exposure to high-pressure steam followed by explosive decompression, disrupts the matrix structure and liberates fibers from bound minerals and contaminants. The modified material demonstrates substantially improved separation characteristics in conventional mechanical processes, achieving higher recovery rates with reduced chemical inputs. Similar benefits have been observed with microwave-assisted extraction systems, where selective heating of polar molecules creates localized pressure differentials that disrupt binding between different sludge components.

Phased implementation strategies for hybrid systems enable mills to manage investment requirements while progressively enhancing recovery rates. Initial deployment typically focuses on primary recovery of the most readily accessible and valuable component—typically fiber for primary sludge or biogas for secondary sludge—with provision for subsequent addition of complementary technologies addressing residual streams.

This approach allows operational experience with the primary recovery system to inform optimal design of secondary processes while generating immediate economic benefits that can partially fund subsequent implementation phases.

Technical challenges for hybrid systems include managing interfaces between different process technologies, particularly regarding consistency adjustments and contaminant profiles that may affect downstream processes.

Successful implementations incorporate buffer storage capacity with adjustable consistency control to accommodate varying throughput rates across different unit operations. Water management presents particular complexity in integrated systems, requiring careful mass balance analysis and potential inclusion of intermediate water treatment stages to prevent contaminant accumulation in recirculated process streams.

2.5.3 Industry 4.0 and Digital Tools for Sludge Management

The progressive digitalization of industrial processes encompassed within the Industry 4.0 paradigm offers significant potential for enhancing sludge valorization through improved monitoring, advanced process control, and data-driven optimization. These digital capabilities address several fundamental challenges in sludge processing, particularly the high variability of input characteristics and complex interdependencies between process parameters and product qualities.

Advanced monitoring systems employing real-time sensors overcome limitations of traditional laboratory-based characterization methods that provide only intermittent data with significant time delays. Online fiber morphology analyzers employing image processing algorithms continuously assess key parameters, including fiber length distribution, fines content, and contaminant profiles, enabling immediate process adjustments to maintain optimal separation conditions. Similarly, mineral content monitors utilizing X-ray fluorescence provide continuous composition data for inorganic fractions, critical for processes requiring specific chemical dosing relative to mineral content. These monitoring capabilities transform sludge processing from a relatively blind operation to a precisely controlled process responding dynamically to input variations.

Predictive modeling tools leverage accumulated operational data to anticipate process outcomes under varying conditions, enabling proactive parameter adjustments rather than reactive corrections after quality deviations. Machine learning approaches have demonstrated particular efficacy for complex sludge processes where traditional first-principles models struggle to capture the multifaceted interactions between biological, chemical, and physical phenomena. Models trained on historical process data can predict recovery efficiency, product quality parameters, and energy consumption based on measurable sludge characteristics and operating conditions, supporting both real-time control and strategic planning decisions regarding sludge management pathways.

Digital twins—virtual replicas of physical recovery systems that simulate behavior under varying conditions—provide powerful tools for process optimization without disrupting ongoing operations. These computational platforms enable evaluation of alternative operating strategies, identification of efficiency bottlenecks, and virtual testing of process modifications before physical implementation. The approach has demonstrated particular value for complex hybrid systems where interactions between different unit operations create non-intuitive response patterns difficult to optimize through conventional methods. As these digital models mature through continuous calibration against actual operational data, their predictive accuracy progressively improves, enabling increasingly sophisticated optimization approaches.

Supply chain integration through digital platforms creates new possibilities for sludge valorization beyond individual mill boundaries. Online marketplaces connecting sludge generators with potential users of recovered materials reduce transaction costs and information asymmetries that have historically limited industrial symbiosis development. Blockchain-based traceability systems address concerns regarding recovered material provenance and quality assurance, providing transparent documentation throughout multi-stage valorization chains. These digital market

TABLE 2.3
Digital Technologies' Impact on Sludge Management Performance Indicators

Technology Category	Implementation Complexity	Capital Investment	Operational Impact	Typical ROI Timeframe
Online monitoring	Medium	Medium	High	1–2 years
Predictive modeling	High	Low-medium	Medium-high	1–3 years
Digital twins	Very high	Medium	Medium	2–4 years
Supply chain platforms	Medium	Low	Medium	1–2 years
AR maintenance support	Medium	Medium	Medium	2–3 years

mechanisms expand the potential application scope for recovered materials by efficiently connecting specialized recovery operations with diverse end-users across different industrial sectors. Operational support systems leveraging augmented reality technologies enhance maintenance and troubleshooting capabilities for increasingly complex recovery systems.

Maintenance personnel equipped with Argumented Reality enabled devices receive contextual information overlaid on their physical view of equipment, including operating parameters, maintenance histories, and step-by-step procedure guidance. This capability proves particularly valuable for sludge processing equipment prone to fouling or performance degradation, enabling preemptive interventions based on early detection of developing issues through integrated condition monitoring systems.

Integration challenges for digital technologies include ensuring data quality and consistency across diverse systems, managing cybersecurity risks associated with increased connectivity, and developing workforce capabilities to effectively utilize advanced digital tools.

Successful implementations typically follow phased approaches where initial digitalization focuses on core monitoring capabilities that generate immediate operational benefits while establishing the data foundation for more sophisticated applications, including predictive modeling and optimization. This evolutionary approach enables progressive capability development while generating continuous improvement in recovery system performance.

2.6 CONCLUSION

Paper mill sludge, once viewed primarily as a waste management challenge, increasingly represents an opportunity for material and value recovery within evolving circular economy frameworks. The technological pathways explored throughout this chapter demonstrate that systematic approaches to sludge valorization can transform this industrial byproduct from environmental liability to valuable resource, contributing simultaneously to economic and sustainability objectives. The progression from conventional disposal methods toward integrated recovery systems reflects broader industrial evolution toward closed-loop material cycles and more efficient resource utilization.

The recovery of fibers and fillers from sludge through mechanical, chemical, and hybrid approaches provides direct replacement for virgin materials within papermaking processes, reducing both raw material costs and environmental footprints. The demonstrated technical feasibility of achieving product quality suitable for various paper grades, particularly in packaging applications, establishes practical pathways for material recirculation within the industry. While quality limitations compared to virgin materials necessitate thoughtful application selection and potential blending approaches, the fundamental viability of these recovered components for papermaking has been conclusively established through numerous commercial implementations.

Utilization of sludge components in paperboard production represents a particularly promising application domain where the specific properties of recovered materials can be advantageously leveraged while accommodating their limitations. The enhanced internal bonding characteristics and improved formation properties contribute positively to certain board performance attributes, while strategic product positioning can mitigate the impact of potential strength reductions or increased moisture sensitivity.

Economic assessment consistently demonstrates attractive returns on investment, particularly in regulatory environments with progressive waste management policies that internalize environmental costs of conventional disposal.

Sustainability analysis through life cycle assessment methodologies confirms the environmental advantages of recovery pathways compared to traditional disposal routes. The substantial reductions in greenhouse gas emissions, fossil resource consumption, and land use impacts establish sludge valorization as an important component of industry decarbonization strategies. These quantifiable environmental benefits strengthen the business case for recovery technologies, particularly as carbon pricing mechanisms and other policy instruments increasingly monetize environmental performance.

The emerging technologies and hybrid approaches discussed in the final sections point toward further advancements in recovery efficiency and economic viability. Integration of mechanical, biological, and thermal processes enables more comprehensive value extraction from heterogeneous sludge streams, while digital technologies provide unprecedented capabilities for process monitoring and optimization. These developments suggest that the frontier of feasible recovery will continue expanding, progressively reducing the fraction of sludge requiring final disposal and increasing the value generated from this material stream.

Future research priorities should focus on addressing remaining technical challenges including enhanced separation of contaminants, improved energy efficiency of recovery processes, and development of novel applications for recovered materials beyond traditional paper products. Regulatory frameworks will need continued evolution to facilitate material circulation while maintaining appropriate environmental and health safeguards, particularly regarding potential contaminants of emerging concern. Collaborative approaches involving paper manufacturers, technology providers, research institutions, and regulatory authorities offer the most promising pathway for advancing sludge valorization practices from current best-available technologies toward truly sustainable circular systems.

The transformation of paper mill sludge from waste to resource exemplifies a broader industrial transition toward circular economy principles, demonstrating practical implementation of closed material loops within a traditional manufacturing sector.

As resource constraints and environmental imperatives continue intensifying globally, the approaches outlined in this chapter provide valuable guidance for other industries seeking similar transformations of their waste streams, contributing to the broader sustainable development agenda while simultaneously enhancing economic performance.

BIBLIOGRAPHY

Monte, M. C., Fuente, E., Blanco, A., & Negro, C. (2009). Waste management from pulp and paper production in the European Union. *Waste Management*, *29*(1), 293–308.

Likon, M., & Trebše, P. (2012). Recent advances in paper mill sludge management. *Industrial Waste*, 73–90.

Mäkelä, M., Watkins, G., Pöykiö, R., Nurmesniemi, H., & Dahl, O. (2012). Utilization of steel, pulp and paper industry solid residues in forest soil amendment: Relevant physicochemical properties and heavy metal availability. *Journal of Hazardous Materials*, *207*, 21–27.

Simão, L., Hotza, D., Raupp-Pereira, F., Labrincha, J. A., & Montedo, O. R. (2018). Wastes from pulp and paper mills: A review of generation and recycling alternatives. *Cerâmica*, *64*, 443–453.

Gupta, G. K., & Shukla, P. (2020). Insights into resource generation from pulp and paper industry wastes: Challenges, perspectives, and innovations. *Bioresource Technology*, *297*, 122496.

Pati, R. K., Vrat, P., & Kumar, P. (2006). Economic analysis of paper recycling vis-à-vis wood as raw material. *International Journal of Production Economics*, *103*(2), 489–508.

Han, Q. L., Zhao, H. L., Wei, G. X., Zhu, Y. W., Li, T., Xu, M., Guo, X., Shi, H. Z., Lian, Y., & Liu, H. Q. (2024). Sustainable papermaking in China: Assessing provincial economic and environmental performance of pulping technologies. *ACS Sustainable Chemistry & Engineering*, *12*(11), 4517–4529.

Faubert, P., Barnabé, S., Bouchard, S., Côté, R., & Villeneuve, C. (2016). Pulp and paper mill sludge management practices: Challenges in assessing impacts on greenhouse gas emissions. *Resources, Conservation and Recycling*, *108*, 107–133.

de Alda, J. A. (2008). Feasibility of recycling pulp and paper mill sludge in the paper and board industries. *Resources, Conservation and Recycling*, *52*(7), 965–972.

Geng, X., Zhang, S. Y., & Deng, J. (2007). Characteristics of paper mill sludge and its utilization for the manufacture of medium density fiberboard. *Wood and Fiber Science*, 345–351.

Tofani, G., De Nys, J., Cornet, I., & Tavernier, S. (2021). Alternative filler recovery from paper waste stream. *Waste and Biomass Valorization*, *12*(1), 503–514.

3 Valorization of Paper Mill Sludge for Sustainable Building Construction Materials

Saakshy

3.1 INTRODUCTION

The construction industry consumes 32% of global energy and contributes 34% of global carbon dioxide emissions. The construction industry is highly dependent on cement and steel, which contribute to 18% of global emissions (www.unep.org). The demand for energy and natural resources has multiplied due to urbanization, population growth, and ongoing improvements in living standards (Vo et al. 2024). Due to a shortage of natural resources, the construction and cement industries are under pressure to meet the massive demand for concrete. Furthermore, environmental contamination has reached concerning levels in both rich and emerging nations as a result of industrial revolution advancements. The disposal of solid, gaseous, and liquid effluents from the construction and cement industries is a matter of concern since they are harming the environment and endangering human and animal life. Approximately 10% of global carbon dioxide emissions are caused in the manufacturing of cement (Bildirici 2019). The carbon footprint of the cement industry is a matter of concern and has led the industry to adopt sustainable materials. The waste from the cement industry can alter the pH of the soil, affecting its ability to sustain plant growth. Large tracts of fertile land have been rendered barren and unproductive for agriculture due to careless disposal practices from the construction and cement industry.

According to the United Nations Environment Program (UNEP) and Global Alliance for Building and Construction (Global ABC), the related climate goals have emphasized carbon reduction by 2030 through initiatives: (1) climate actions roadmap, (2) building breakthrough, (3) energy efficiency, (4) material carbon, and (5) policy implementation. The sustainable building challenges are given in Figure 3.1.

The solid waste of the paper industry is produced primarily from the rejects of screening, primary and secondary sludge from wastewater management, and lime sludge from the chemical recovery system. Per estimates, about 40–50 kg of dry sludge is generated in the production of 1 ton of paper at a paper mill, out of which approximately 70% is primary sludge and 30% is secondary sludge.

DOI: 10.1201/9781003461395-3

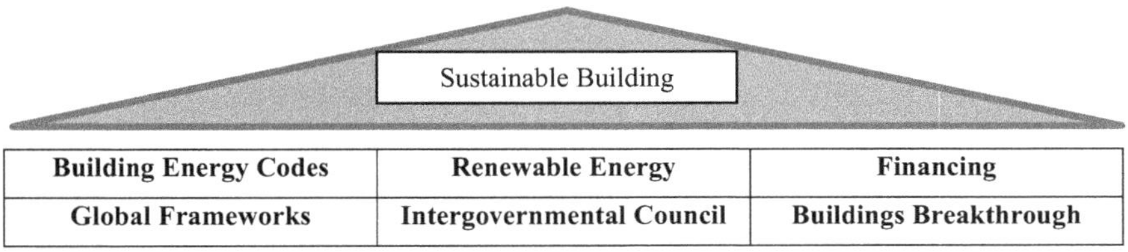

FIGURE 3.1 Sustainable building challenges.

The sludge generated from the paper industry is one of the major waste types generated from the paper industry. The generated waste from the pulp and paper industry causes severe harm to aquatic life, disturbs the food chain, and also has various health implications. The quantum of disposal of paper industry waste is a matter of concern, and reuse of the waste to generate useful resources is a big step towards green energy and environmental sustainability.

3.2 CLASSIFICATION OF PAPER MILL SLUDGE

3.2.1 Primary Sludge

Primary sludge in the paper industry is the solid residue recovered from the initial clarification of wastewater during paper production. It primarily consists of cellulose fibers, fillers (like kaolin and calcium carbonate), and other additives that were incompletely separated from the wastewater during the pulping and papermaking processes.

3.2.2 Primary Sludge of the Paper Industry

3.2.3 Destination Sludge

"Destination sludge" in the paper industry refers to the sludge generated by deinking mills during the deinking process of recycled paper fibers, which primarily consists of ink, fillers, and residual fibers.

3.2.4 Secondary Sludge

Secondary sludge in the paper industry is a by-product of the secondary wastewater treatment process, often referred to as activated sludge. It is the sludge that remains after biological treatment of paper mill wastewater, primarily composed of microbial cell mass and organic matter.

3.2.5 Combined Sludge

Combined sludge in the paper industry refers to a mixture of primary and secondary sludge generated during wastewater treatment processes in pulp and paper mills. It's

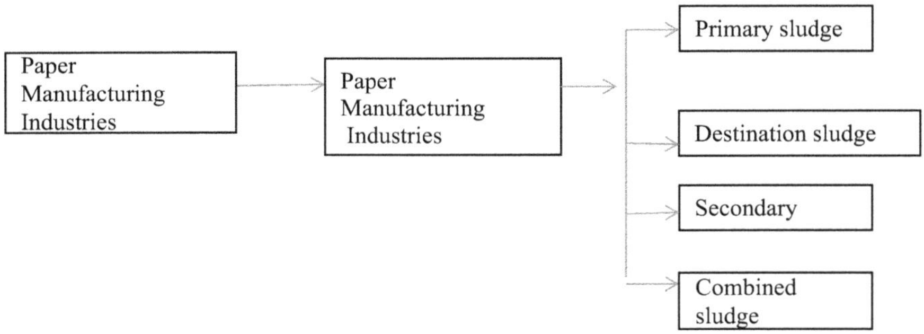

FIGURE 3.2 Classification of paper manufacturing waste.

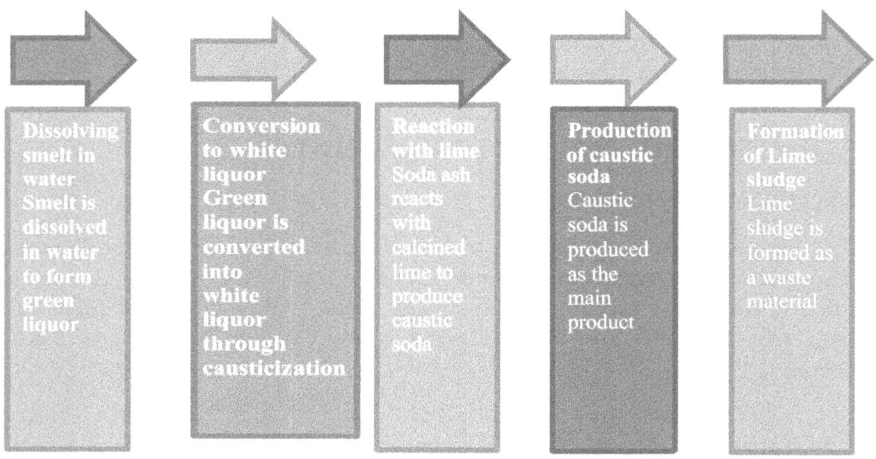

FIGURE 3.3 Formation of lime sludge.

a by-product of the papermaking process that includes cellulose fibers, fillers, and other materials.

3.2.5.1 Paper Mill Lime Sludge

Lime sludge is generated from the chemical recovery section of the paper industry. Approximately 1.6 tons of paper mill lime sludge is generated for 1 ton of paper. The process of generation of paper mill lime sludge is given in Figure 3.3.

The paper industry faces problems in the disposal of lime sludge, as lime sludge has high silica content and creates scaling problems in the kilns if recycled in lime kilns. Therefore, land application and landfilling are the only options left for paper industries, but due to stringent regulations and the scarcity of open land, the utilization of paper mill lime sludge has been explored as an optimum option. About up to 30% of lime sludge can be used in concrete and mortar. Lime sludge can also be used in the production of cement and ceramics (Vashishtha et al. 2019).

3.2.5.2 Paper Sludge Ash

Sludge from the paper industry is incinerated to reduce its volume and also as a source of fuel. Paper sludge ash is created in amounts of 10–15 kg per ton of paper. Paper sludge ash is a type of industrial waste produced by the pulp and paper industry that poses both environmental and financial difficulties. In the field of construction engineering, paper sludge ash has been used as an additional cementitious component in the creation of concrete. Its pozzolanic qualities increase the strength and durability of concrete. Some studies have explored the use of super-hydrophobic paper sludge ash (PSA) to improve concrete's resistance to water ingress.

3.3 CHARACTERIZATION OF SLUDGE FROM THE PAPER INDUSTRY

The characterization of sludge from the paper industry is given in Table 3.1. The characterization of sludge from the paper industry depends on the raw material, process adopted, and chemicals used in the identified paper industry, but there are still some common chemicals, like calcium oxide; kaolin; and major amounts of degraded fiber, which is grayish in color. Paper industry sludge shows pozzolanic activity and cementitious properties.

3.3.1 Organic Matter

The organic matter content in paper mill sludge varies from 32.5 to 58%, of which almost 25% is cellulosic fibers.

3.3.2 Elemental Analysis of Paper Mill Sludge

Paper mill sludge is mainly composed of carbon (49.3%), oxygen (30.4%), hydrogen (6%), nitrogen (3.2%), and 11.1% other inorganic matter.

TABLE 3.1
Characterization of Primary Sludge

S.N.	Parameters	Goel et al. (2017)	Primary Sludge De Azevedo et al. (2019)	Gimenez et al. (2009)	Raw Hypo Sludge Source: CEP	Portland Cement
1	Calcium oxide, CaO,%	5.37	70.84	19.82	46.2	50–60
3	Silica, SiO_2,%	14.56	8.03	18.05	9.0	20–25
4	Magnesium oxide, MgO,%	5.96	–	2.58	3.3	2–3
5	Iron oxide, Fe_2O_3,%	2.06	0.65	0.55	–	1–2
6	Aluminum oxide, Al_2O_3,%	10.02	9.46	10.14	–	5–10

TABLE 3.2
Significance of Constituents of Sludge From the Paper Industry in the Construction Industry

S.N.	Component in Sludge	Significance	References
1	Calcite in primary sludge	Promotes the formation of vitreous stage during sintering when incorporated in a ceramic material	Buruberri et al. (2015)
2	Talc	Used in ceramic industry	De Azevedo et al. (2019) Bhushan et al. (1954)

TABLE 3.3
Characteristics of Portland Cement

S.N.	Particulars	Portland Cement Source: Velumani et al. (2020) [11]
1	Specific gravity	3.12
2	Fineness	8%
3	Initial settling time, minutes	29
4	Consistency,%	33

3.3.3 XRF Characterization of Paper Sludge

Analysis of the chemical composition by XRF confirms that the principal oxides are calcium oxides, SiO_2, and Al_2O_3.

3.3.4 Evaluation of Paper Waste Samples by XRD and FESEM

According to Vashishtha et al., lime sludge showed the presence of calcite, talc, and kaolinite. Schneider et al. provided an overview considering all the possibilities of lime sludge use in the production of cement. Calcination of lime sludge produces metakaolin ($Al_2O_3\_2SiO_2$) that is highly reactive and can be used as an additive material in cement

The high content of calcium oxide promotes the formation of the vitreous phase during sintering, which helps in decreasing thermal conductivity. At around 700°C, kaolin converts into amorphous meta-kaolinite. Paper sludge can also be a source of both calcium and sodium silicate. It contains calcium-rich compounds like calcium carbonate and calcium oxide as well as silicate phases.

Portland cement showed the following characteristics, as shown in Table 3.3.

The initial settling time of cement varies between 30 to 60 minutes, and after the addition of paper mill sludge, the settling time was reduced (Turner et al. 2022).

TABLE 3.4
Amount of Generated Paper Sludge

Mill Type	Kraft Mill	Sulfite Mill	Deinking Mill
Amount of generated sludge, kg/T	58	102	234

Source: Scott (1995)

3.4 QUANTIFICATION OF PAPER SLUDGE

Paper sludge contains 40–60% dry matter with 35–40% calcite, cellulose, calcium carbonate, and a small amount of organic compounds and heavy metals. The amount of sludge generated from the paper industry varies according to the raw material and pulping processes.

From Table 3.4, it is quite clear that a deinking mill using waste paper or recycled fiber generates the maximum amount of sludge due to shorter fibers.

3.5 UTILIZATION OF PAPER SLUDGE IN THE CONSTRUCTION INDUSTRY

3.5.1 Paper Sludge in Cement

The inorganic content of paper sludge, such as calcium carbonate, can be used as raw material in cement and brick manufacturing. Paper sludge can also act as a secondary source of calcium carbonate (calcite) in one part geopolymer cements. The calcium compounds present in paper sludge can be used as an alternative to lime or cement. The organic content in paper sludge enhances water retention. The presence of silicates in paper sludge ash makes it suitable for use in geopolymer concrete, where it can act as a partial replacement for fly ash or other pozzolans (Hassan et al. 2025).

3.5.2 Paper Sludge in Tiles

The study conducted by Alqam et al. (2011) reported that tiles, except for 50% sludge cement replacement, comply with the breaking strength requirements of 2.8 MPa for tiles for external applications. Alqam et al. investigated the use of paper sludge to replace cement in the production of paving tiles for outdoor use, which produced tiles with a water absorption ratio of around 10%. Research confirms that using paper sludge (or its ash) content (around 10–20%) to partially replace cement in outdoor paving tiles is feasible, with studies showing around 10% water absorption and meeting standards for non-vitreous tiles. These eco-friendly tiles offer sustainable waste management, though proper processing of the sludge (drying, sometimes incineration/adding silica fume) is crucial for performance.

3.5.3 Paper Sludge in Bricks

India makes about 250 billion bricks per year and is the second-largest producer of brick after China. Fired bricks are traditionally used due to their physical, mechanical, and thermal properties; compressive strength; durability; and compactness. Bricks consume a quantum amount of soil, and many countries have started the reduction of the use of fertile soil for making bricks. Dried and powdered PMS can be incorporated into brick manufacturing, potentially reducing the need for traditional clay-based bricks. Paperboard sludge admixed with pozzolanic and cementitious materials has been used for making bricks in which 20% by weight of cement has been used to make modified bricks. The paper sludge up to maximum 20% was utilized to produce eco-friendly bricks that meet the requirements for second-class bricks as per Bureau of Indian Standards. The compressive strength and flexural strength of conventional concrete increases up to 90 days with compressive strength often reaching near its ultimate capacity around 90–99% by 28 days. Lightweight casting floor tiles were prepared by mixing fly ash, bentonite, and cement materials (Goel & Kalamdhad 2017; Velumani & Manikandan 2020; Ravikumar et al. 2025).

The physical and mechanical properties of the mixture after the addition of paper mill sludge per BIS methods and durability studies to meet requirements include compressive strength, flexural strength, water absorption, curing time, and dry density. The bricks are classified per compressive strength from 3.5 to 35.0 N/mm^2.

3.5.4 Paper Sludge in Concrete Mixtures

The sludge from paper manufacturing industry is utilized as a replacement to the mineral fillers in various concrete mixes. Concrete mixes containing various contents of waste were prepared, and basic strength characteristics, such as compressive strength, splitting, flexural strength, water absorption, and density, were determined and compared with a control mix. Five concrete mixes containing various contents of waste, 0 (control mix), 3, 5, 8, and 10%, as a replacement for fine sand were prepared with ratios of 1:3:6 by weight of cement, sand, and aggregate, respectively. The test results revealed that as the content of waste increased, the water-to-cement ratio for the mix also increased, since the waste had a high degree of water absorption. Therefore, additional water was required for cement hydration. The results obtained showed that as the amount of waste increased, the basic strengths, such as compressive strength, decreased. A maximum of 5% content of waste as a replacement for fine sand in concrete mixes can be used successfully as construction materials, such as in concrete masonry construction, with a compressive strength of 8 MPa, splitting strength of 1.3 MPa, water absorption of 11.9%, and density of 20 KN/m^3.

Paper sludge was utilized as a replacement for mineral fillers in concrete mixes for non-structural masonry construction. The presence of waste material reduced the concrete density. There was an increase in the water absorption of the concrete mixes as the content of waste increased.

A compressive strength of 8 MPa, which was achieved by using 5% waste, is acceptable for use in masonry construction. A reduction in splitting tensile strength was observed when using different replacement percentages of waste. Waste from paper manufacturing has very high water absorption. Concrete mixes subjected to compaction pressure with as much as 20% waste produced higher compressive and splitting tensile strengths.

- Lime sludge application in cement mortars
- Application in bricks and ceramics
- Application of lime sludge in other building materials
- Environmental impact advantages of lime sludge application in building materials

Paper mill sludge can be used as a partial replacement for cement in concrete and mortar mixes. This reduces the amount of cement needed, decreasing manufacturing costs and carbon emissions.

Research suggests PMS can be used to create lightweight concrete products, reducing the overall weight of structures (Ahmadi & Al-Khaja 2001). The important properties of concrete include compressive strength, flexural strength, and water absorption. Paper mill sludge creates lightweight concrete products, reducing the overall weight of structures.

3.6 CONCLUSION

Paper sludge, after mixing with cement, proved suitable for pozzolanic material, showing positive effects on settling times and an improvement in consistency. The workability, mechanical characteristics, durability, and cost feasibility of a sustainability analysis of paper sludge improved the characteristics of concrete. Scanning electron microscopy showed the formation of calcium silicate hydrate, which improved bonding in concrete. X-ray diffraction analysis showed the crystalline structure of paper sludge in concrete mixtures. Studies conducted have found that up to 10% paper sludge can be utilized in concrete. The specific composition of PMS can vary, so proper testing and analysis are needed to determine its suitability for various applications. The utilization of paper mill sludge not only reduces waste disposal issues and promotes environmental sustainability but also lowers the cost of construction materials. Thus, they are comparatively less expensive and an alternative to traditional building materials.

REFERENCES

Ahmadi, B., & Al-Khaja W. (2001). Utilization of paper waste sludge in the building construction industry. Resource Conservation & Recycling, 32, 105–113.

Alqam, M., Jamrah, A., & Daghlas, H. (2011). Utilization of cement incorporated with water treatment sludge. Jordan Journal of Civil Engineering, 5(2), 268–277.

Bhushan, B., & Roy, H. N. (1954). Occurrence of talc in India and its role in ceramic industry. Transactions of the Indian Ceramic Society, 13(4), 211–220. https://doi.org/10.1080/0371750X.1954.10877663

Bildirici, M. E (2019). Cement production, environmental pollution, and economic growth: Evidence from China and USA. Clean Technologies and Environmental Policy, 21(4), 783–793.

Buruberri, L. H., Seabra, M. P., & Labrincha, J. A. (2015). Preparation of clinker from paper pulp industry wastes. Journal of Hazardous Materials, 286, 252–260.

De Azevedo Afonso, R. G., Alexandre, J., Pessanha, L. S. P., Manhales, R. S. T., Brito, J. D., & Marvila, M. T. (2019). Characterizing the paper industry sludge for environmentally safe disposal. Waste Management, 95, 43–52.

Gimenez, R.G., Vegas, I., & De La Villa, R.V. (2009). Ion mobilisation and transport through cement mortars blended with thermally activated paper sludge in natural climatic conditions. Water Air Soil Pollution, 203, 39–52.

Goel, G., & Kalamdhad, A. S. (2017) An investigation of paper mill sludge in brick manufacturing. Construction and Building Materials, 148, 334–343.

Hassan, M. L., Hassan, E. A., Elseoud, W. S. A., & Moustafa, A. M. (2025). Utilization of paper sludge in preparation of high purity calcium formate. Biomass Conversion and Biorefinery, 15, 4565–4578.

Jwaida, Z., Dulaimi, A., Alyhya, W., Algretawee, H., & Al-Busaltan, S. (2024). Recycling and utilization of paper sludge ash-current status review and future perspectives. Sustainable Materials and Technologies, 40.

Ravikumar, B. S., Chandra, H. S. S., & Surendra, H. J. (2025). Waste paper sludge ash as a pozzolanic material: Enhancing concrete performance and sustainability. Journal of Building Pathology and Rehabilitation, 10(3). https://doi.org/10.1007/s41024-024-00518-z

Turner, T., Wheeler, R., & Oliver, I. W. (2022). Evaluating land application of pulp and paper mill sludge: A review. Journal of Environmental Management, 317, 15439.

Vashishtha, P., Kumar, V., Singh, S. K., Dutt, D., Tomar, G., & Yadav, P. (2019). Valorization of paper mill lime sludge via application in building construction materials: A review. Construction and Building Materials, 211, 371–382.

Velumani, P., & Manikandan, P. (2020). Contribution to the development of tiles made of paper board sludge. Advances in Environmental Research, 9(4), 285–294.

Vo, D. H., Vo, A. T., & Ho, C. M. (2024). Urbanization and renewable energy consumption in the emerging ASEAN markets: A comparison between short and long run effect. Heliyon, 10, e30243.

www.unep.org

4 Coal Fly Ash to Pigment and Filler for Papermaking

Suresh Sundaramurthy, Anuj Kumar Verma, and Arisutha Suresh

4.1 INTRODUCTION

4.1.1 COAL: AN EXTENSIVE RESOURCE

The growth of any country depends on effective conversion technology to turn industrial waste materials, such as fly ash, into valuable products. The lack of suitable utilization strategies causes global environmental problems. Mining and other mineral exploration processes are increasing the demand for natural resources and destabilizing ecosystems (Sinha et al. 2008). Because there is not 100% utilization of fly ash in India or the majority of other countries, some of the fly ash is disposed of in nearby places, such as power plants, causing pollution by mixing with air, water, and soil (Tiwari et al. 2016). Normally, silica, alumina, and iron oxides are the major components of fly ash. In thermal power plants, fly ash is a byproduct of burning coal, and electrostatic separators are used to collect it (Bayat 2002).

The worldwide available coal is over 850 gigatons, which is sufficient for more than 130 years at the current rate of production (WCA 2012). The highest quantity of coal is available in Russia, China, North America, Australia, and India.

4.1.2 CONSUMPTION AND UTILIZATION OF FLY ASH IN INDIA AND THE WORLD

In 1996, the generation of fly ash in India was 68.88 million tons, whereas the consumption was 9.65% fly ash. In 2016, the generation of fly ash increased to 107.10 million tons, while the consumption also jumped to 63.28%. This amounts to 33% of fly ash still being unused and causing many environmental problems (Singh 2015).

Table 4.1 shows power generation, coal consumption, and ash generation in India.

In an effort to utilize 100% of the fly ash generated from power plants and various other sources, the government of India has been continuously providing short-term training and projects to increase the usage of fly ash in various sectors, such as cement, agricultural, bricks and tiles, mine filling, and concrete, but such efforts have not yet achieved full utilization of fly ash in India (Nawaz, 2013). Table 4.2 shows the consumption of fly ash in various sectors in India.

DOI: 10.1201/9781003461395-4

TABLE 4.1
Power Generation, Coal Consumption and Ash Generation in India (Senapati 2011)

Year	Power Generation (mW)	Coal Consumption (million tons)	Ash Generation (million tons)
2000	75,000	250	95
2010	100,000	300	115
2020	138,000	400	145

TABLE 4.2
Modes of Fly Ash Utilization in Various Industries in India

S. N.	Mode of Utilization	Million Tons	Percentage (%)
1	Cement	40.5869	23.98
2	Mine filling	11.7827	6.96
3	Brick and tiles	14.9110	8.81
4	Reclamation of low-lying areas	11.0392	6.52
5	Ash dyke raising	11.8888	7.02
6	Roads and flyovers	6.1942	3.66
7	Agriculture	1.9243	1.14
8	Concrete	0.7647	0.45
9	Hydro power sector	0.0197	0.01
10	Others	7.984	4.72
11	Unutilized fly ash	62.1577	36.72
	Total	**169.2534**	**100**

Source: Government of India (Singh 2015)

Coal has been a dominant player in energy generation worldwide. Fly ash generation and utilization is highest in China, consuming almost half of the global coal supply. India shows the largest growth in consumption, overtaking the US to become the world's second-largest consumer of coal (Sarkar et al. 2005).

From the data in 2010, as shown in Table 4.3, globally, the production of coal combustion products (CCPs) was approximately 780 million metric tons (Mt). The highest CCP-producing countries were China at 395 Mt, North America at 118 Mt, India at 105 Mt, Europe (EU15) at 52.6 Mt, and Africa 31.1 at Mt, with the Middle East as a minor contributor. Australia contributes approximately 2% to global production of CCPs.

From the 780 Mt produced, some 415 Mt, or 53%, were reported as utilized. Utilization varies widely in the countries discussed in this chapter. Japan had the highest reported effective utilization rate of 96.4%, and Africa/the Middle East had the

TABLE 4.3
Typical Range of Elemental Composition for CCPs From Different Coals, wt.% (Barnes 2010; Heidrich et al. 2007)

Country/ Region	CCPs Production (Mt)	CCPs Utilization (Mt)	Utilization Rate%	CCPs Production/ Person (Mt)	CCPs Utilization/ Person (Mt)
Australia	13.1	6.0	45.8%	0.60	0.27
Canada	6.8	2.3	33.8%	0.20	0.07
China*	395.0	265	67.1%	0.29	0.20
Europe (EU15)	52.6	47.8	90.9%	0.11	0.10
India*	105.0	14.5	13.8%	0.09	0.01
Japan	11.1	10.7	96.4%	0.09	0.08
Middle East and Africa	32.2	3.4	10.6%	0.02	0.01
Other Asia*	16.7	11.1	66.5%	0.05	0.03
Russian Federation	26.6	5.0	18.8%	0.19	0.04
Total/s	**777.1**	**415.5**	**53.5%**		

Note: *Non-members of World Wide Coal Combustion Products Network (WWCCPN).

lowest at 10.5%. Countries ranked with the highest coal combustion product utilization rates were: Japan with 96.4%, Europe with 90.9%, China with 67%, and other Asian countries with 66%. Australia's coal combustion product utilization rate was 45.9%, or just below the global average of 53%.

4.1.3 Importance of Coating Pigments and Fillers in the Paper Industry

Papermakers have started looking at all significant components of paper products. The two major components of any grade of paper are pulp and fillers. In recent years, high filler content has become important in the paper industry to reduce overall production costs. The lower cost of fillers compared to virgin fiber can result in significant savings in raw materials and considerable cost reductions in the drying section of paper machines.

The addition of filler can improve paper optical properties, sheet formation, dimensional stability, printability, and appearance (Yuan 2008; Shen et al. 2007; Lu et al. 2010). However, filler increases have several disadvantages and limitations (Han 2009; Shen et al. 2009) such as fiber bonding, low retention, high RDF dosages, and low bulk. To respond to the current business situation around the world, there is an acute need to introduce innovation in filler engineering to meet these industry challenges. At present, developing new fillers and incorporating them with fiber to enhance paper properties while lowering energy consumption and fiber costs is a major area of interest.

4.2 FLY ASH AS A COATING PIGMENT AND FILLER IN THE PAPER INDUSTRY

Fly ash is a byproduct of various industrial processes, and its conversion into valuable products, such as fillers and pigments, can enhance paper properties and serve as a substitute for fiber in writing, printing, and other paper grades. Since fly ash has a porous structure, it can provide better bulk and porosity as compared to conventional pigments and fillers such as PCC, GCC, kaolin clay, TiO_2, and talc (Sinha 2008). Using calcium silicate filler made from fly ash in a high-pressure carbonation method (Mathur 2001) to achieve fiber substitution and other surface properties for writing, printing and other grades of paper manufacturing is a suitable utilization technique to increase the consumption of fly ash. The critical need for the paper industry can be summarized as reducing the costs of manufacturing, water, power, and energy consumption per unit weight of paper by improving profitability through value-added products such as fly ash–based calcium silicate. There is also an urgent need to address environmental issues due to fly ash generated from power plants, as nearly 35% of this material is unused. Papermakers must consider alternative materials and methods to get better paper quality at low cost.

In the paper industry, finely divided white mineral pigment as a coating material provides special properties to paper such as surface gloss, weight, and reduced ink absorbency (Andersson 2008). The pigment fills the voids between the fibers and coating itself and thus improves the properties of the sheet (Chinga and Helle 2003). The most common papermaking pigments are kaolin clay (typical composition of 39% Al_2O_3, 46% SiO_2, and 13% H_2O), calcium carbonate (98 to 100% $CaCO_3$), titanium dioxide (98 to 100% TiO_2), and hydrated silica (78% SiO_2, 5% CaO, and 17% H_2O). Kaolin clays are commonly used pigment materials in papermaking, but the cost is high. Paper printability and appearance can be increased with the help of mineral coating on any grade of paper (Laufmann et al. 2000). Various properties of the end product depend upon pigment, binders, and their amounts in a coating formulation.

When two different types of pigment are used for the same coating weight, paper properties depend upon pigment shape, size, its distribution, and aggregation. Since pigment is a major component, it must be analyzed in an effort to achieve better optical, surface, and printing properties (Preston et al. 2001). It has been stated that "seven S factors" play vital roles in coating; these are size, shape, spacing, surface, spread, stirring, and structure (Patton 1979). Cost is another factor that motivates replacement of expensive fiber by coating, as pigment is a major component in coating formulations (Kettle et al. 2010). Fly ash has the same enhancement opportunities available as these common pigments (Song et al. 2012), but there no research on the application of fly ash as a coating pigment in papermaking is available.

4.2.1 Filler and Its Types in the Paper Industry

Filler is not just to fill the space between fiber and fines; it impact the paper structure, surface properties, and many more valuable properties, especially optical properties. Filler also increases the friction and its interaction with inks (Biza 1999). Filler dimension varies from 0.2 to 4 μm. Fiber length varies from 1 to 3 mm, whereas filler length is 1/1000 of the length of fiber. Due to their small particle size, the retention of fillers plays a crucial role in papermaking because they serve as a replacement for expensive fiber (Schwalbe 1962; Gill and Hagemeyer 1992; Scott 1996). Filler presence is more often tested by simple laboratory techniques like ash analysis of the paper sample and measurement of strength and optical properties. Talc is the most dominant filler in the papermaking, but the quality of the paper is low (Gill 2004). PCC and GCC become the major filler pigments as compared to talc by providing excellent paper properties. Another reason carbonate is preferred over talc as a filler is the change in process conditions from acidic to alkaline papermaking. Six to 8 million tons per year of mineral filler and pigment are used by papermakers worldwide. Pulverized coal fly ash is used as a filler in laminated grades of paper (Sinha, 2008). The strength properties of laminated grades of paper can be increased using fly ash–based filler, but optical properties have high deviation, so fly ash can be used where brightness and other optical properties are not required. High-brightness fly ash filler can be produced, so fly ash–based filler can be used for a variety of paper grades (Song 2008). In 2013, Zhang suggested that fly ash can be a good filler where whiteness is required and showed FPAR and FPR can be improved using FACS filler and that it can be a potential filler for writing printing paper. Still, the optical and strength property enhancement actually required for writing printing paper is not reported using fly ash as a filler to the authors' knowledge.

4.2.1.1 Precipitated Calcium Carbonate

In 1980, talc was the dominant filler used in the paper industry, whereas precipitated calcium carbonate was growing. At present, PCC is the most widely used filler material worldwide. In the USA, PCC use increased nearly 70% (Gill 2004). Precipitated calcium carbonate also saw increases in Asia, Europe, and other parts of the world (Tøttrup 1999). The cost of opacity is the motivation for the growing demand for PCC in the paper industry (Williams and Gill 1992).

Before 1980, TiO_2 was used to increase opacity for lightweight paper, but it was very costly as compared to other fillers present at the time. Later, due to the high light-scattering coefficient of PCC, papermakers were able to get the desired opacity (Gill 1988). Limestone is the raw material used for the manufacturing of precipitated calcium carbonate. The first step is calcination, in which CaO is produced. The second step is precipitation, in which water is added to CaO (Laine 1980). The reaction of water with calcium oxide produces calcium hydroxide. The further reaction of $Ca(OH)_2$ with carbon dioxide gives precipitated calcium carbonate, which is insoluble in water (Hagemeyer 1984).

Various shapes and sizes can be produced depending on the temperature and pressure conditions. Peng et al. (2015) reported that flocculating lime mud before adding it to the fiber slurry resulted in improved strength properties compared to untreated lime mud (Peng et al., 2014). The removal of Zn and Sr from PCC mineral increases the light-scattering efficiency, as reported in the literature (Koivunen et al. 2009 and Koivunen and Paulapuro 2010). Treated PCC with silica (Lourenço et al. 2013 and 2014) contributes better strength properties. An SEM image of PCC is shown in Figure 4.1.

FIGURE 4.1 SEM image of PCC.

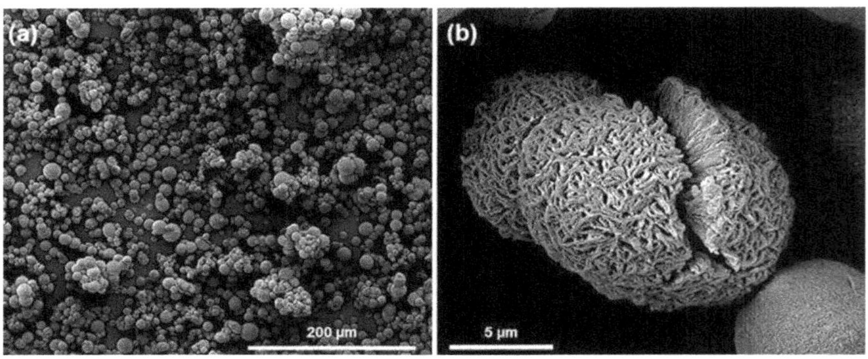

FIGURE 4.2 SEM picture at various magnifications (Pinkl et al. 2018).

4.2.1.2 Urea-Formaldehyde Pigment—A Non-Mineral Filler

Using the condensation product of urea and formaldehyde, a non-mineral filler can be produced with greater optical properties (Menashi et al. 1978; Kramer 1993). The SEM image at various magnifications is given in Figure 4.2.

Due to the high light scattering coefficient, void structure, and bulking properties, urea formaldehyde is considered a good filler for papermaking (Kramer and Haller 1993). The refractive index of urea formaldehyde pigment is lower, but its light-scattering ability is as high as that of TiO_2 (Kramer 1993).

4.2.1.3 Starch as Filler

Starch can be added in papermaking either in liquid form or in granular form depending on the process conditions. Krogerus's 1999 research on starch as a filler material in papermaking shows an increase in opacity when starch is used in a granular form. Shen et al. (2011) show starch as an organic filler used in papermaking. Kim et al.

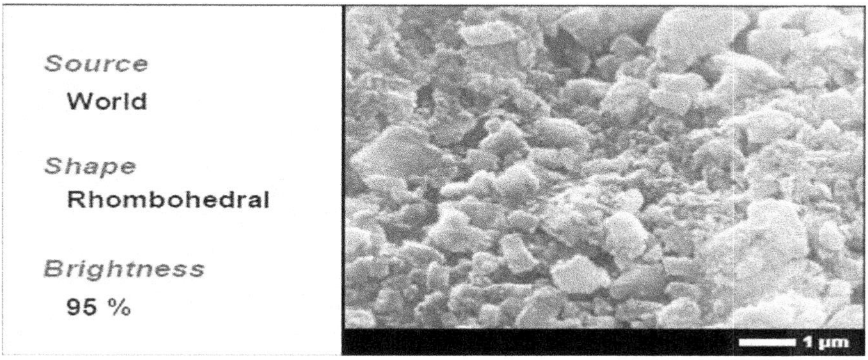

FIGURE 4.3 GCC, marble based (Omya 1998).

(2015) used various powders supporting lignocellulosic material as a filler in papermaking. Results show that starch can be used as a coating material on filler and fiber to increase optical, surface, and strength properties.

4.2.1.4 Ground Calcium Carbonate

Limestone deposits are the main source of ground calcium carbonate or ground limestone (Hagemeyer 1984; Huggenberger and Neubold 1997). Limestone with fewer impurities is available in India. The Moh's number of hard limestone is in the range of 3 to 4 (Wypych 1999). Mining is the first step to release the material from the rock formation. After mining, grinding is done; this plays an important role in getting different particle sizes. Polyacrylate dispersants and phosphate are added during the grinding process to help in the processing of the minerals. The negative surface charge of GCC negatively impacts wet-end papermaking performance, leading to reduced first-pass ash retention and total first-pass retention (Gill and Hagemayer 1983). There are two size classes of GCC: fine ground (70%, <2 μm) and ultrafine ground (90%, <2 μm) (Gill and Hagemeyer 1983). Furthermore, these large particles are removed by fractionation in order to remove abrasiveness on wire.

GCC has become a major choice for papermakers due to its high brightness and low cost; moreover, it is very helpful in drainage and drying energy savings. (Laufmann et al. 2000). An SEM image of GCC is given in Figure 4.3.

4.2.1.5 Kaolin Clay

In ancient times, kaolin clay was mined in China, so its name is based on a hill in China. The term "China clay" is used for kaolin clay, while kaolinite used for the pure mineral, which is obtained from sedimentary and metamorphic deposits (Sennett 1997). In the early 1900s, clay was the preferred filler used in the wet end of papermaking (Harben and Virta 1999). Major parts of the world, such as Georgia and South Carolina in the USA, Cornwall and parts of Wales in the United Kingdom, and Brazil, have high purity clay deposits. In the USA, it is mined by excavation and then mixed by high shear mixer. The aspect ratios of kaolin clay are in the range of 4 to

12 (Slepetys and Cleland 1993; Sennett 1997; Beckett et al. 1997). The process steps involved in clays are screening, fractionation, bleaching, and magnetic removal of impurities. Some colored impurities are removed by flotation during the processing of clays (Jameson and Sennett 1989). The most important properties that make kaolin clay a filler for the production of paper are brightness, fewer abrasives, and ease of dispersal in water when phosphate is used as a dispersing medium (Solomon and Hawthorne 1983). The quality of clay depends upon the source and degree of process (Velho and Gomes 1995). Darkening of the pulp is reported in some of the literature when iron is present in the clay (Narvestad et al. 2013).

SEM images of US and European kaolin are given in Figures 4.4 and 4.5.

4.2.1.6 Calcined Clay

A brighter, bulkier filler material can be produced by heating hydrous kaolin material in a high temperature furnace. (Drzal et al. 1983; Dubois et al. 1995). A similar treatment can be applied to talc or aluminum trihydrate to get higher brightness and

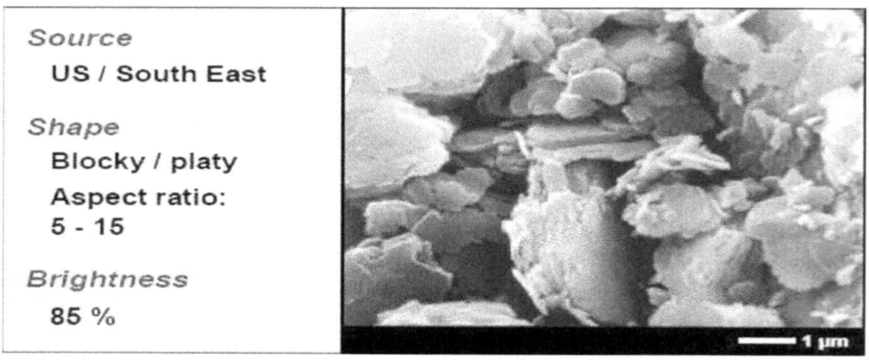

FIGURE 4.4 US kaolin (Omya 1998).

FIGURE 4.5 European kaolin (Omya 1998).

bulkier material (Mahapatra and Patel 1994). Calcination means the conversion of calcium carbonate to calcium oxide at a temperature above 950°C (Drzal et al. 1983). A high light scattering coefficient and large area of air-to-solid interface make calcined clay a good choice to use as a filler material for papermaking (Bauch 1992). An image of calcined clay lumps is given in Figure 4.6.

4.2.1.7 Talc

In order to control pitch and stickies, talc is most widely used. (Trivedi 1997). However, talc is also used as a filler material in papermaking all around the world (Biza 1999). For stickies and pitch problems, talc is used as a detackifier at a dosage of 5 to 7.5 kg per ton of paper. (Rooks 1993). Talc Moh's number is 1. Talc is known for its softness, oil-loving character, and platy particle shape (Chauhan and Bhardwaj 2013). The surface free energy of talc is in the range of 35 to 68 m/NM. Talc particles decrease friction and increase various paper properties (Biza 1999). An SEM image of European talc is given in Figure 4.7.

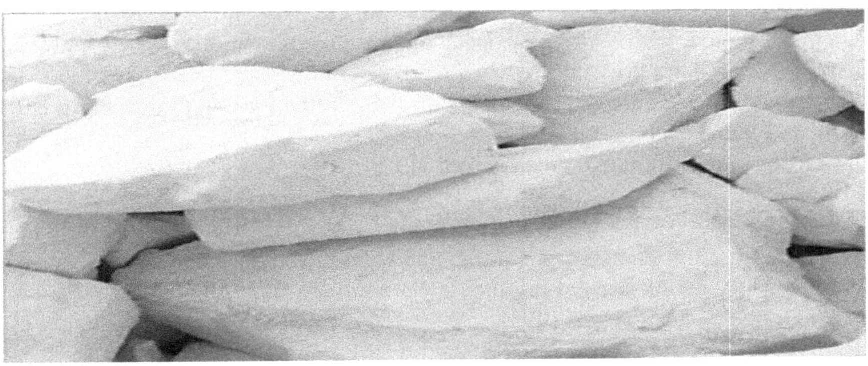

FIGURE 4.6 Calcined clay lumps.

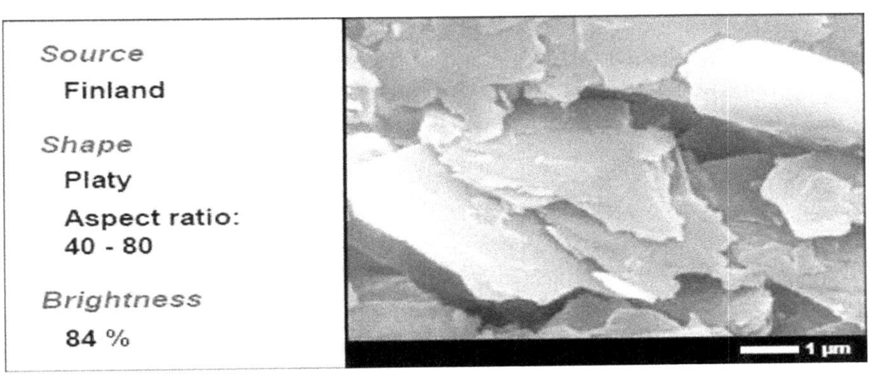

FIGURE 4.7 European talc (Omya 1998).

FIGURE 4.8 Mica lumps.

4.2.1.8 Mica

Sericite mica is found mostly in Taiwan and is used as paper filler due to its availability in that region. It has the ability to swell and resembles bentonite (Perng and Wang 2004). It can act like a microparticle in the wet end of the papermaking process. Mica as a filler with flocculants and coagulants increases retention and drainage properties (Perng and Wang 2004).

An SEM image of mica lumps is given in Figure 4.8.

4.2.1.9 Diatomaceous Earth

It is a rarely used filler in papermaking, and it mainly consists of silica. Due to its highly abrasive nature, desired brightness, and porosity, it is suitable to be used as a papermaking filler (Cho et al. 1995).

4.2.1.10 Calcium Silicate

Mathur (2006) tested the filler (quartz/dolomite) on papermaking in a high-pressure batch reactor. All the properties (mechanical/optical/surface) of paper sheets were improved. However, these fillers are very expensive. Some scientist investigated fly ash/$Ca(OH)_2$ (fly ash-to-calcium hydroxide ratio) on papermaking in a reactor operated at 90°C. They found that a 3:1 ratio of fly ash/$Ca(OH)_2$ with a surface area of 9.77 m^2/g improved whiteness from 29.8 to 66.63% ISO. Sinha et al. (2008) used fly ash as a filler for decorative paper, where brightness and whiteness are not so important, and showed fly ash as a filler can improve strength properties but decreases optical properties due to the grey-black color of fly ash. Zhang et al. (2013) briefly investigated the filler and fiber combined refinery process to investigate the effect on various paper properties. The ratio used was 0.15–0.5%. In a mixed refinery, the bulk reduced from 2.04 to 2.85 cm^3/g, while the traditional method was from 2.5 to 3.24 cm^3/g. However, optical and mechanical properties were increased. Zhang et al. (2013) tested FACS filler in papermaking, where the bulk increased by 56.4%, whiteness increased to 91.5%, and surface area was 121 m^2/g. They used a pre-flocculation technique for filler increase in paper ranging from 12 to 25%. However, only mechanical properties increased, and optical properties were not so improved. Wang et al. (2015) used fly ash–based calcium silicate as a filler in papermaking and he

FIGURE 4.9 SEM image of calcium silicate (Zhange et al. 2013).

FIGURE 4.10 Powder of gypsum as a filler for papermaking.

investigated the effect of mass ratio of calcium carbonate to fly ash in this ratio for making calcium silicate (1:0, 4:1, 2:1, 1:1, 1:2, 1:4, 0:1).

They found that a 1:1 ratio was optimal, with a 74% whiteness improvement compared to raw fly ash material (only 30.3% ISO). An SEM image of calcium silicate is given in Figure 4.9.

4.2.1.11 Gypsum

Gypsum (calcium sulfate dehydrate) is the cheapest among all fillers available on the market. Gypsum can be prepared in various shapes and sizes as a filler to be used in various types of paper machines. The most used form of the crystal is a needle-like shape with a length-to-diameter ratio of about 4 to 10 (Gussinyer 1986). From a profitability point of view, gypsum can be prepared with positive zeta potential, which is useful for good retention. On the other hand, the main drawback of gypsum is solubility with water; nearly 2 to 4 grams of gypsum per liter of water will dissolve during papermaking, resulting in an increased load on effluent treatment plants (ETPs) (Hofmann et al. 1990). Gussinyer and Gussinyer (1999) proposed a mechanism whereby evaporation of a saturated solution during the drying of paper tends to fuse the ends of gypsum crystals onto the adjacent cellulosic fiber, creating a permanent bond. Powder of gypsum as a filler is shown in Figure 4.10.

4.2.1.12 Amorphous Silicate Fillers

The chemistry of synthetic amorphous silicate is similar to that of kaolin clay, with a higher surface area per unit mass. Due to its high porous structure, per Koga and Taga (1993), this filler can be very effective for absorbing printing ink. It can also be a very effective coating material for a traditional filler (Novak 1982). This material can be very useful to control the particle morphology used to get better paper properties. An SEM image of amorphous silicate fillers is given in Figure 4.11.

4.2.1.13 Aluminum Tri-Hydroxide

It is also called ATH and is used in the production of paper where high brightness is required (Healy and Kramer 1997; Chen et al. 2011a). It is very small and spherical in shape and used to achieve higher optical properties (Healy and Kramer 1997). It is also used as a fire retardant filler in papermaking (Chen et al. 2011). It is the best substitute for TiO_2, which is very costly as compared to other filler material available in the paper industry (Koenig and Tauth 1984). Scaling is the biggest drawback of this filler (Healy and Kramer 1997). The solid structure of aluminum tri-hydroxide is given in Figure 4.12.

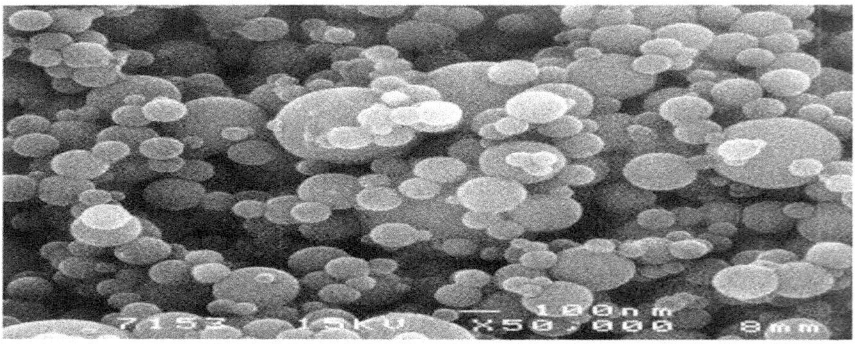

FIGURE 4.11 SEM of amorphous silicate fillers (www.denka.co.jp).

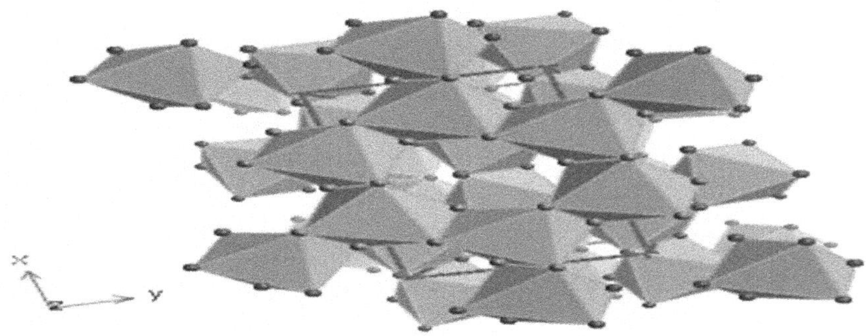

FIGURE 4.12 Solid state structure of aluminum tri-hydroxide.

4.2.1.14 Titanium Dioxide

Titanium dioxide is the most widely used filler material for high-quality paper that requires high opacity. It has many advantages, but due to its high price and highly abrasive nature, papermakers are searching for new filler materials for the paper industry (Kwoka 1997). The chemical formula of titanium dioxide is TiO_2. There are three mineral forms available for titanium dioxide depending upon the method of synthesis (Solomon and Hawthorne 1983). These three forms are brookite, rutile, and anatase, but only anatase and rutile are used as papermaking fillers in the paper industry. Both are tetragonal in shape. The particle sizes of TiO_2 are in the range of 0.25–0.45 μm (Wypych 1999), and the light scattering coefficient is about 1035 m^2/kg. An image of titanium dioxide is given in Figure 4.13.

4.2.1.15 Acid-Tolerant PCC

Calcium carbonate filler causes pulp to yellow or darken, so new filler is required to overcome this issue, such as acid-tolerant PCC (Poirier et al. 1996). Calcium carbonate filler dissolves in water when the pH is below 6.5, so it is recommended to avoid acid papermaking if calcium carbonate is used as a filler, but in the alkaline region, pulp loses its brightness. Since brightness is a selling factor in the paper industry, both parameters must be considered simultaneously. An SEM image of acid-tolerant PPC is given in Figure 4.14.

To overcome this limitation (Mathur and Lasmarias 1991; Evans et al. 1991; Ain and Laleg 1997; Shen et al. 2009a,b,c), various authors have suggested acid-tolerant PCC, which gives satisfactory results without compromising brightness or sizing efficiency (Patel and Panigrahy 1996a,b). Acid-tolerant PCC was manufactured by various authors. In 2003, Pang et al. suggested that $CaCO_3$ dissolution in water can be prevented if it is treated with phosphate. Sufficient time provided for the layer of calcium phosphate results in no dissolution in water. In 1990, Tokarz et al. gave a process where sodium silicate solutions were added to PCC slurry, followed by the pH being adjusted to 9.8 and the addition of $ZnCl_2$.

FIGURE 4.13 Titanium dioxide (Omya 1998).

FIGURE 4.14 SEM of acid-tolerant PCC (Lourenço et al., 2016).

4.2.1.16 Aragonite

Aragonite has the same chemical formula as precipitated calcium carbonate, but the arrangement of atoms in the crystal is different (Laine 1980; Dunlop-Jones and Jaycock 1981; Anderson and Yunko 1983; Hagemeyer 1984; Konno et al. 2007; Hu et al. 2009). It requires higher temperatures and is sometimes referred to as needles. Its length-to-thickness ratio is 5. It is mainly used in coating applications as a pigment. Hu et al. (2009) used aragonite as filler, and their result showed greater strength and optical properties. Wang et al. (2012) showed that the rod-like structure of aragonite can be prepared during the processing of lime mud. Chen et al. (2011) showed high retention and strength properties using long aragonite crystal instead of ordinary precipitated calcium carbonate for 10% filler replacement. Using the caustic zing property of conventional kraft pulp mill, a high quality of aragonite crystal can be prepared (Konno et al., 2007 and Nanri et al. 2008).

4.2.1.17 Chalk

Chalk is a porous, white, soft, sedimentary carbonate rock. It is a form of limestone composed of mineral calcite (Dunlop-Jones and Jaycock 1981). England, France, Denmark, and Belgium are the major countries in which good-quality chalk is found (Dunlop-Jones and Jaycock 1981). An SEM image of chalk is given in Figure 4.15.

Chalk has been used as a filler in papermaking since 1960, but product brightness was not acceptable (Burke 1993). Chalk also has less effect on strength properties as compared to other material available as paper filler. Less grinding energy is required to grind chalk as compared to limestone. The size of chalk particles is predefined and depends upon the shell-forming process. The surface area of chalk particles is in the range of about 3 to 5 m^2/g (Fairchild and Clark 1996), which is much less than the present filler available for papermaking. A paper requiring high bulk and more stiffness has different process conditions for filler than paper requiring more opacity and smoothness. There are four major papermaking fillers derived from natural mined materials, and these and their properties are shown in Table 4.4.

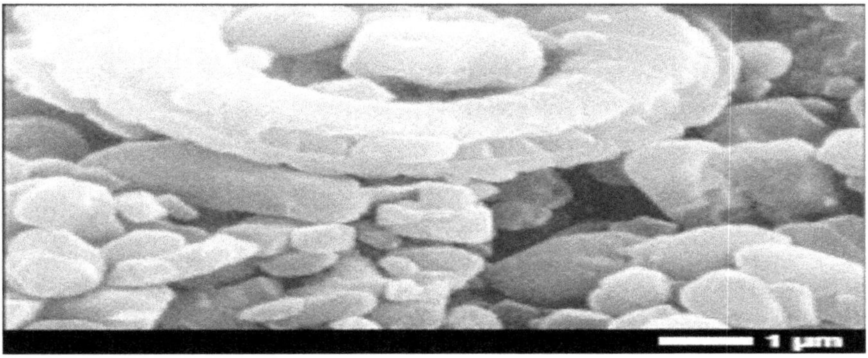

FIGURE 4.15 Chalk brightness 86% (Mollaahmad, 2008).

TABLE 4.4
Typical Properties of Four Major Papermaking Fillers Derived From Natural Mined Materials (Hubbe and Gill 2016)

Product Type	Clay (Kaolin)	GCC (Limestone)	Chalk	Talc
Density (g/cm^3)	2.6	2.7 to 2.9	2.7 to 2.9	2.7 to 2.85
Hardness (Moh's scale)	2	3 to 4	3 to 4	1 to 1.5
Loss on ignition, 900°C (%)	12 to 14	43.5	43.5	4.8 to 17
Refractive index	1.55	1.48, 1.65, 1.7	1.48, 1.65, 1.7	1.58
Mean diameter (μm)	2 to 5	0.8 to 2	2	1 to 20
Surface area (m^2/g)	10 to 20	5 to 12	3 to 5	2.6 to 35
Typical particle shape(s)	Platy, bookettes	Irregular, blocky	Coccolith, rounded	Platy
Aspect ratio (length/width)	4 to 30	1 to 1.5	1 to 1.5	5 to 20
Reflectivity (%, 547-nm light)	80 to 90	95	88	78 to 93
Abrasion (Einleiner)	2 to 5	4 to 8	–	3 to 5

There are four major types of synthetic and precipitated filler, as shown in Table 4.5.

Since olden days, papermakers have used a variety of fillers for various grades of paper to reduce the overall cost of paper manufacturing. There are some major benefits as well as some drawbacks, per a literature review (Schwalbe 1962; Kapoor et al. 1996). Some of the most important benefits are brightness, opacity, dimensional stability, smoothness, print quality, and above all the overall cost of the product. Tables 4.6 and 4.7 present the advantages and disadvantages of filler material used in paper.

TABLE 4.5
Typical Values of Four Types of Major Synthetic Fillers Used in the Paper Industry (Gill and Hagemeyer 1983; Wypych 1999)

Product Type	PCC (Calcite)	TiO_2 (Rutile)	$Al(OH)_3$ (ATH)	$CaSO_4.2H_2O$
Density (g/cm^3)	2.7–2.9	4.24	2.4	2.3–3
Hardness (Moh's scale)	3–4	6–7	2.5–3.5	2–3
Loss on ignition (%)	43.5	0–2	34.6	Various
Refractive index	1.48, 1.65, 1.7	2.7	1.58	1.52–1.61
Mean diameter (μm)	0.8–2	0.3–0.35	0.5–1	6
Surface area (m^2/g)	4–25	–	8–16	–
Typical particle shape(s)	Rosette blocky	Rounded	Rounded	Needle
Aspect ratio (length/width)	1–1.5	1–1.5	1–1.5	10
Reflectivity (547-nm light)	98–99	98	99–100	90–92
Abrasion (Einleiner)	3 to 5	30	1 to 3	6 to 8

TABLE 4.6
Advantages of Filler Used in Paper (Hubbe and Gill 2016)

Benefits	Ways to Reduce Cost and Improve Paper Properties	Citations
Cost reduction	Increasing filler contents means reducing fiber	Davidson 1965
Increase in brightness	Brightness is the selling factor of paper, and adding a brighter filler will increase the brightness of the paper	Bauch 1992
Increase in opacity	High refractive index filler utilization in paper-making means an increase in opacity	Bown 1998
Reduction in drying energy	Increasing filler content in paper means a decrease in drying energy in the dryer section	Kapoor et al. 1996
Friction increase	Scalenohedral PCC filler or pigment	Rooks 1993
Friction decrease	Talc as a filler	Biza 1999
Pore size	Size and shape of filler play important roles in papermaking fillers	Schwalbe 1962
For cigarette paper, a controlled burning rate	Aluminum tri-hydrate (ATH) uses can reduce	Schwalbe 1962

Very few literature reviews are available for fly ash as a filler for papermaking. There is no literature review available for fly ash as a pigment for newsprint paper property measurement to date. Based on a literature review, Sinha et al. (2008, 2010, 2011) utilized fly ash as a filler material for decorative specialty paper, where brightness and whiteness did not play a vital role. They showed that fly ash can improve the strength properties, but optical properties were very low. Fan and Qian (2012)

TABLE 4.7
Disadvantages of Filler Used in Paper (Hubbe and Gill 2016)

Related Problem	Ways to Minimize Problem	Citations
Decrease in strength results in a large number of breaks on the paper machine	Increase pulp refining and utilization of starch, and control particle morphology can reduce web breaks and control the strength	Bown 1985b
Paper apparent density enhancement	In order to control the apparent density issue, high bulk filler is the most suitable	Moberg 1985a, 1985b
Low FPAR and two-sidedness issue	Coagulation and flocculation must be very effective to get better retention and drainage properties	Tanaka et al. 1982
In forming fabric, abrasiveness problems	Avoid hard and fine filler material	Laufmann and Rapp 1995
Dusting, contamination, and stickiness problem	Starch with a high degree of substitution (DS) should be used to avoid this problem	Fineman and Hoc 1978
Sizing chemical demand	Using less filler and larger particle sizes can be a probable solution to the sizing problem	Krogerus 1999; Chabot et al. 2004

utilized cationic polyacrylamide (cPAM) as an aid to increase the retention of fly ash material as a filler in papermaking. Since the brightness of fly ash was low, they utilized a floatation process, a screening process, and a combination of fly ash with precipitated calcium carbonate to get the desired level of brightness. Results showed a significant improvement in brightness, but as filler dosage increases (more than 10%), brightness decreases. Overall, the strength properties of fly ash–filled sheets increase, but optical properties decrease. Zhang et al. (2013) tried fly ash as a filler material for papermaking. The result shows that fly ash–based calcium silicate has extra properties such as high bulk, high surface area (121 m^2/g), and larger particle size as compared to conventional filler such as PCC. Retention and drainage were better than with conventional filler, whether using cPAM or not. This is due to the larger particle size, which prevents filler particles from passing through the wire section of the paper machine. Bulk improvement (56.4%) was observed using fly ash as a filler material, and other strength properties was superior as compared to PCC, whereas optical and brightness properties were lower (nearly 2 to 5 points), as seen in the previous literature (Fan and Qian 2012). The chemical and physical properties of various types of fly ash obtained from different courtiers show similar properties, which are comparable with the chemical and physical properties of kaolin clay, precipitated calcium carbonate, and other pigments used in the paper industry. Figure 4.16 shows the important physical and chemical properties of fly ash.

Scanning electron microscopy analysis has shown that fly ash particles have irregular shapes, with hollow spherical pores (Shreya et al. 2014). Energy dispersive X-ray analysis (EDAX) has revealed the presence of a high amount of silicate

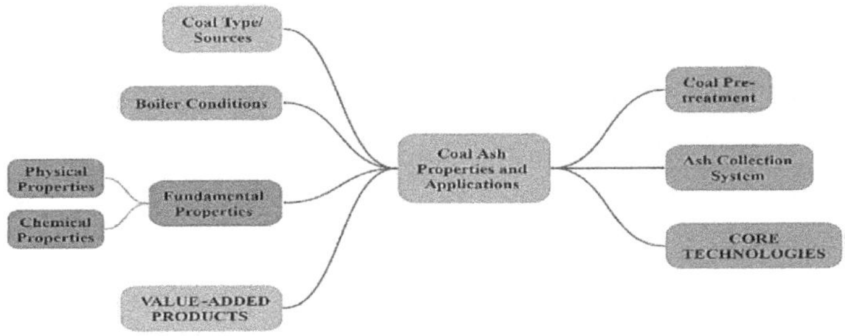

FIGURE 4.16 Important physical and chemical properties of fly ash (Jow et al. 2015).

FIGURE 4.17 Typical fly ash as seen with SEM.

Source: Adapted from Garg, 2016.

and mineral oxides with amorphous natures (Shreya et al. 2014). The major constituents are mullite and quartz. Lime, magnetite, hematite, gypsum, and rutile were the other minerals present in minor amounts (Smichowski et al. 2008). The addition of fly ash–based calcium silicate pigment for coating to the newsprint base sheet changes its pore structure and enhances its properties, such as brightness, stiffness, breaking length, and tear index. Fly ash addition to the newsprint base sheet increases the opacity of the paper remarkably and better than kaolin clay (Sinha 2008). However, the application of fly ash as filler and pigment in the newsprint industry has not yet been reported. An SEM image of typical fly ash is shown in Figure 4.17.

4.3 SUMMARY

Results obtained from different researchers for newsprint paper quality (fly ash as a pigment material) showed low-cost waste fly ash can be used as a coating pigment in newsprint for quality improvement, which provides economic and

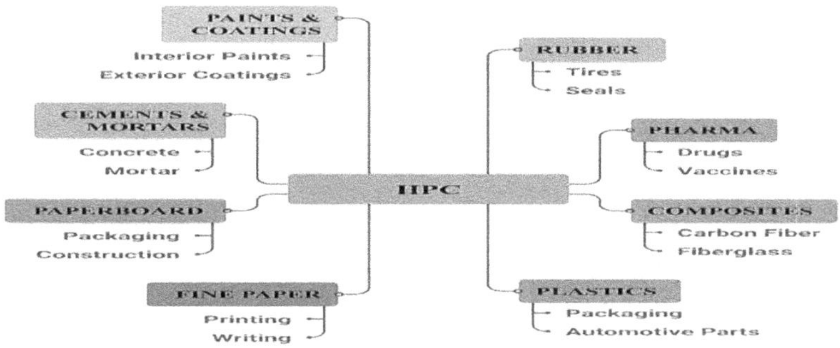

FIGURE 4.18 Future scope of present research work.

environmental benefits. The most significant increase was in brightness, which generates the potential for higher revenue in newsprint. Newsprint made with a calcium silicate pigment coating can compete with imported newsprint. Fly ash generated from the power plant in a newsprint mill can be used as a potential coating pigment in the size press for energy savings during drying and result in higher opacity of the newsprint and superior mechanical strength properties. The production of modified fly ash–based coating pigment using a high-pressure carbonation process shows excellent optical and strength properties when compared to PCC and the base newsprint sheet.

For writing printing paper, results show that fly ash–based filler has unique properties as compared to PCC due to high pressure carbonation and pre-flocculation techniques. Modified FACS have high FPR and FPAR as compared to unmodified PCC with low starch consumption. Modified FACS shows a slight drop in opacity and brightness as compared to unmodified PCC, as treated filler have a much less negative impact on optical properties. Bulk reduction is a direct consequence of higher filler loading, as reported in various studies, but the reflocculation technique has a much lower or negligible effect on bulk. FACS filled sheet bulk increases contrary to various research data indicating that increasing ash decreases the bulk of the sheet. Predefined filler size, floc strength, and its control distribution reduce the negative impact on strength properties, contrary to literature stating that on increasing filler, strength properties decrease.

Drying and refining energy will automatic decrease as filler loading increases. Results also show that FACS filler can be sized properly as desired by the papermaker. Fillers are used in different application areas such as paper, rubber, plastic, adhesives, coatings, and sealants. Applications of filler are countless, and approximately 53 million tons of fillers are used in different sectors. FACS can be a potential filler, which can singlehandedly be a substitute for a variety of applications due to its chemical and physical nature. Fly ash–based calcium silicate can be used in various industries, including for plastic, fine paper, paperboard, paint and coating, cement, glass, rubber, and composite, as a filler material. The future scope of present research work is given in Figure 4.18.

BIBLIOGRAPHY

Sinha, A. S. K. (2008). Effects of pulverized coal fly-ash addition as a wet-end filler in papermaking. *TAPPI Journal*, *7*(9), 3–7.

Bayat, B. (2002). Comparative study of adsorption properties of Turkish fly ashes: The case of nickel (II), copper (II), and zinc (II). *Journal of Hazardous Materials*, *95*(3), 251–273.

World Coal Association. (2012). *World Coal Association report* (Vol. 2012).

Singh, M. (2015). *Report on fly ash generation at coal/lignite-based thermal power stations and its utilization in India (2014–15)*. Central Electricity Board, New Delhi.

Nawaz, I. (2013). Disposal and utilization of fly ash to protect the environment. *International Journal of Innovative Research in Science, Engineering and Technology*, *2*(10), 5259–5266.

Senapati, M. R. (2011). Fly ash from thermal power plants: Waste management and overview. *Current Science*, 1791–1794.

Sarkar, A., Rano, R., Mishra, K., & Sinha, I. (2005). Particle size distribution profile of Indian fly ash: A comparative study. *Fuel Processing Technology*, *86*(11), 1221–1238.

Barnes, I. (2010). Ash utilisation: Impact of recent changes in power generation practices. In *IEA Clean Coal Centre Report* (Vol. 1, p. 51).

Yuan, J. H. (2008). *A new fiber loading technology for high ash content paper* (PhD thesis). Nanjing Forestry University.

Shen, J., Song, Z. Q., & Qian, X. R. (2007). Strength improvement of paper containing calcium carbonate filler. *Journal of China Pulp and Paper*, *26*(5), 47–51.

Han, C. (2009). *Starch-coated PCC in papermaking* (Master's thesis). Nanjing Forestry University.

Shen, J., Song, Z. Q., Qian, X. R., & Liu, W. (2009). Modification of papermaking fillers: A review. *BioResources*, *4*(3), 1190–1209.

Mathur, V. K. (2001). High-speed manufacturing process for precipitated calcium carbonate. *U.S. Patent No. 6, 251, 356.*

Andersson, C. (2008). Enhancing functionality of paperboard by surface treatment: A review. *Packaging Technology and Science*, *21*(6), 339–373.

Chinga, G., & Helle, T. (2003). Coating surface variation and print quality. *Journal of Pulp and Paper Science*, *29*(6), 179–184.

Laufmann, M., Forsblom, M., Strutz, M. D., & Yeakey, S. R. (2000). GCC vs. PCC as paper fillers. *TAPPI Journal*, *83*(5), 76.

Preston, J., Nutbeem, C., Parsons, J., & Jones, A. (2001). Printability of coated papers. *Paper Technology*, *42*(2), 33–41.

Patton, T. C. (1979). *Paint flow and pigment dispersion*. Wiley.

Kettle, J., Lamminmäki, T., & Gane, P. (2010). Modified surfaces for inkjet coating. *Surface and Coatings Technology*, *204*(12–13), 2103–2109.

Song, S., et al. (2012). Fly ash-based calcium silicate filler. *Industrial & Engineering Chemistry Research*, *51*(50), 16377–16384.

Biza, P. (1999). Talc as a multifunctional mineral. In *Minerals in Papermaking Conference Proceedings*.

Schwalbe, H. C. (1962). Fillers and loading. In C. E. Libby (Ed.), *Pulp and Paper Science and Technology* (Vol. 2, pp. 60–73). McGraw-Hill.

Williams, W. R., & Gill, R. A. (1992). Precipitated calcium carbonate as filler. *Wochenblatt für Papierfabrikation*, *120*(17), 669–670.

Scott, W. E. (1996). *Principles of wet end chemistry*. TAPPI Press.

Gill, R. A. (2004). Fillers in papermaking. *TAPPI Course Notes*.

Koivunen, K., Alatalo, H., Silenius, P., & Paulapuro, H. (2010). Starch-based fillers. *Journal of Materials Science*, *45*(12), 3184–3189.

Koivunen, K., & Paulapuro, H. (2010). Structured PCC fillers. *TAPPI Journal*, *9*(1), 4–14.

Koivunen, K., Niskanen, I., Peiponen, K. E., & Paulapuro, H. (2009). Nanostructured PCC fillers. *Journal of Materials Science, 44*(2), 477–482.
Lourenço, A. F., Gamelas, J. A. F., & Ferreira, P. J. (2014). Silica-coated PCC fillers. *Nordic Pulp and Paper Research Journal, 29*(2), 240–245.
Menashi, J., Hardy, J. F., Economou, P., Heckman, F. A., & Radman, E. (1978). Urea-formaldehyde pigments. *Journal of Colloid and Interface Science, 66*(2), 220–233.
Kramer, K. (1993). Organic fillers in papermaking. *TAPPI Conference Proceedings.*
Shen, J., Song, Z. Q., Qian, X. R., & Ni, Y. H. (2011). Carbohydrate-based fillers. *Carbohydrate Polymers, 85*(1), 17–22.
Kim, S. Y., Lee, J. Y., Kim, C. H., Lim, G. B., Park, J. H., & Kim, E. H. (2015). Organic filler surface modification. *BioResources, 10*(1), 1174–1185.
Wypych, G. (1999). *Handbook of fillers* (2nd ed.). Plastics Design Library.
Sennett, P. (1997). Kaolin clays. In *Pigments for Paper*. TAPPI Press.
Beckett, R., Murphy, D., Tadjiki, S., Chittleborough, D. J., & Giddings, J. C. (1997). Particle characterization techniques. *Colloids and Surfaces A, 120*, 17–26.
Solomon, D. H., & Hawthorne, D. G. (1983). *Chemistry of pigments and fillers*. Wiley.
Velho, J., & Gomes, C. (1995). Effect of kaolin quality on paper. *Investigação Técnica de Papel, 125*, 585–599.
Narvestad, H., Gregersen, Ø. W., & Kure, K. A. (2013). Clay-induced discoloration. *Nordic Pulp and Paper Research Journal, 28*(1), 68–81.
Mahapatra, S., & Patel, M. (1994). Filler modification effects. *IPPTA Journal, 6*(1), 57–65.
Trivedi, N. (1997). Talcs in papermaking. In *Pigments for Paper*. TAPPI Press.
Chauhan, V. S., & Bhardwaj, N. K. (2013). Talc filler effects. *Appita Journal, 66*(1), 66–72.
Cho, J. H., Yoon, B. H., Cho, B. M., & Lee, J. M. (1995). Diatomaceous earth fillers. *Journal of TAPPI Korea, 27*(4), 56–63.
Zhang, M., Song, S., Wang, J., Sun, J., Li, J. Z., Ni, Y., & Wei, X. (2013). Fly ash-based fillers. *BioResources, 8*(2), 2768–2779.
Zhang, M., Li, Q., Song, S., Hao, N., & Liu, G. (2016). Paper strength improvement using fly ash fillers. *BioResources, 11*(3).
Gussinyer, J. M. (1986). Calcium sulfate fillers. *TAPPI Conference Proceedings.*
Hofmann, H. P., von Raven, A., & Weigl, J. (1990). Gypsum as coating pigment. *TAPPI Journal, 73*(12), 139–147.
Kwoka, R. A. (1997). Titanium dioxide in papermaking. In *Pigments for Paper*. TAPPI Press.
Poirier, N. A., Pikulik, I. I., Labadie, M., & Pageau, A. (1996). Effect of pH on paper properties. *Journal of Pulp and Paper Science, 22*(11), J419–J423.
Evans, D. B., Drummond, D. K., & Koppelman, M. H. (1991). PCC fillers in groundwood papers. *TAPPI Conference Proceedings.*
Burke, D. J. (1993). Evolution of alkaline papermaking. *TAPPI Journal, 76*(4).
Fairchild, G. H., & Clark, E. B. (1996). PCC morphology effects. *APPITA Conference Proceedings.*
Kapoor, S. K., Sood, Y. V., Moorthy, K. S., Pande, P. C., & Rao, N. J. (1996). Increasing filler loading. *IPPTA Journal, 8*(3), 93–99.
Davidson, R. R. (1965). Loading paper with fillers. *Paper Technology, 6*(2), 107–120.
Fan, Y. M., & Qian, X. R. (2012). Fly ash as papermaking filler. *China Pulp and Paper, 31*(4), 22–26.
Moberg, K. (1985). Highly filled papers. *TAPPI Conference Proceedings.*
Tanaka, H., Luner, P., & Côté, W. (1982). Filler distribution in paper. *TAPPI Journal, 65*(4), 95–99.
Chabot, B., Daneault, C., Fournier, F., Morneau, D., & Arial, M. (2004). Retention and drainage methodology. *Paperi Puu, 86*(6), 445–449.
Smichowski, P., Polla, G., Gómez, D., Espinosa, A. J. F., & López, A. C. (2008). Metal fractionation in fly ash. *Fuel, 87*(7), 1249–1258.

5 Decarbonization and the Energy Transition in Waste Management in the Pulp and Paper Industry

Suresh Sundaramurthy and Anuj Kumar Verma

5.1 INTRODUCTION

A high organic waste stream comes from the paper and pulp industry, which is not good for the environment and health. Based on current stringent standards and norms for effluent discharge, all industries are focusing on waste management strategies and proper implementation at the source. The paper and pulp industry is looking for decarbonization of processes and energy-efficient technologies through which a circular economy can be achieved. Biogas via methane production is a good idea for recycling wastes generated from the paper and pulp industry, which can help to reduce CO_2 as well as aid the energy transformation. Several stakeholders have come together to propose green processes for the paper and pulp industry to achieve carbon neutrality. Methane produced by anaerobic digestion can be upgraded, and the upgradation process to produce blended gas, compressed biogas, or bio-compressed natural gas (bio-CNG) is relatively new in the market.

Currently, the paper and pulp industry is upgrading existing processes with energy-efficient technologies, such as switching from fossil fuels to on-site-produced biofuels in combination with combined heat and power (CHP), improving energy management systems, and increasing the use of waste paper and paper recycling. This implementation can be an innovative idea to reduce CO_2 emissions and energy consumption. This chapter deals with methane production, schematic approaches, anaerobic digestion, biogas upgrades, and how this solution can promote carbon neutrality.

5.2 DECARBONIZATION AND THE ENERGY TRANSITION IN WASTE MANAGEMENT IN THE PULP AND PAPER INDUSTRY

Because of huge changes in the climate, all countries need to focus on the pulp and paper sector achieving climate neutrality by 2030/2050. Based on the literature, there are already numerous opportunities for energy transition and decarbonization in the pulp and paper industry, including biogas production through anaerobic digestion

DOI: 10.1201/9781003461395-5

and its upgrade to improve methane quality. In particular, biogas upgrades have been a relatively new process in the market.

The best available technologies and emerging innovative technologies are applicable at different stages of the pulping and papermaking process. For each technology, the energy savings (heat or electricity) and the decarbonization potential in the associated process are considered, and to find out how mature the technology is, the technology readiness level is monitored. For example, biogas can be upgraded to biomethane and injected into the grid for use in hard-to-decarbonize sectors such as transport, where its value is many times higher compared to burning it in a boiler or gas engine.

Biogas is produced by bacteria that decompose organic materials into a blend of gases: 45–85 vol% methane (CH_4) and 25–50 vol% carbon dioxide (CO_2). Apart from CH_4 and CO_2 from the biodigester, other gases released include H_2S, ammonia, and moisture.

Different adsorption towers are used for upgrading biogas into methane, such as sawdust, coconut shell, and activated carbon towers, and for removing trace amounts of gases that lower the quality of methane. Biomethane is the upgraded form of biogas, consisting of almost 100% methane and approximately equal to natural gas in quality. It can also be produced by gasification or power-to-methane technologies.

Biogas can be further purified into methane, which blends with compressed natural gas. Now on the market, there are several blended fuels available such as liquefied natural gas (LNG), biogas-liquefied natural gas (BioLNG), and compressed natural gas (CNG). LNG is natural gas that has been cooled down to liquid form for ease and safety of non-pressurized storage or transport. BioLNG is produced from biomethane. CNG is natural gas compressed to less than 1% of the volume it occupies at standard atmospheric pressure. It is stored and distributed in hard containers at a pressure of 20–25 MPa.

In recent decades, the paper and pulp industrial sector has started to improve efficiency in its processes. High-loading organic waste from the pulp and paper industry can be converted to energy through different conversion technologies such as biochemical and thermal processes. In both processes, energy production takes place in different forms: liquid, gas, and solid. All the forms further convert into compressed natural gas, biogas compressed natural gas, electricity, and heat. In this way, the paper and pulp industry can achieve zero liquid discharge, find substitutes for their feedstocks, and comply with standard norms.

In Austria, the paper and pulp industry has implemented technology integrated with waste heat recovery systems to facilitate on-site biofuel production. This installation has reduced CO_2 by 30% in the past couple of years. In general, most countries target their objectives at controlling the atmospheric CO_2 level through decarbonization, especially in the industrial sectors, by 2050. In the paper and pulp industry, they aim for a reduction of energy usage through different process within their existing technologies by means of choosing energy-intensive steps such as steam cycle washing of pulp and innovative drying processes.

Washing pulp with steam instead of water enhances consistency and reduces energy usage by 30–40%. During the pressing of paper, drying technology is the most important for removing the wet content. The innovative drying technologies found in the literature (in terms of reduction in energy consumption) are impulse (10–25%), condebelt (~15%), gas-fired (10–20%), dry sheet forming, and direct electric heating. Based on reported results in the literature, steam cycle washing, impulse drying, condebelt drying, and direct electric heating are used in the integrated writing paper industry and gas-fired dryer and dry sheet forming in the non-integrated writing paper and integrated tissue paper industry.

5.3 A WAY FORWARD FROM DECARBONIZATION AND THE ENERGY TRANSITION OF THE PULP AND PAPER INDUSTRY WASTE STREAM

Biomethane is chemically equivalent to natural gas and contains the same amount of energy. Biogas or biomethane cogeneration plants allow the simultaneous production of electricity and steam. Biomethane can be used in transport and seasonal energy storage and also sourced using the existing gas networks via biomethane purchase agreements. BioLPG can replace fossil LPG, using the same delivery method.

BIBLIOGRAPHY

Sundaramurthy, S., Singh, S. P., Verma, A. K., Jawaid, M., & Rene, E. R. (Eds.). (2026). Black Liquor Waste Management in the Pulp and Paper Industry (1st ed., Vol. XIV, 190 pp.). Springer.

Ravnjak, D., Plazl, I., & Moze, A. (2007). Kinetics of colloidal alkylketene dimer particle deposition on pulp fibers. Colloid Polym. Sci., 285, 907.

Slepetys, R. A., & Cleland, A. J. (1993). Determination of the shape of kaolin pigment particles. Clay Miner, 28(4), 495–508.

Swerin, A., Sjödin, U., & Ödberg, L. (1993). Flocculation of cellulosic fibre suspensions by model microparticulate retention aid systems. Nordic Pulp and Paper Research Journal, 389–399. U.S. Patent 4,799,964, Jan. 24.

Verma, A. K., Suresh, S., & Mohta, D. (2018). Economical and Efficient Use of Fly Ash for Newsprint Paper Quality Improvement. BioResources, 13(3): 5765–5777.

Wang, J., Wei, P., Liu, P., & Sun, W. (2012). Identifying appropriate conditions for producing spindle-like causticizing precipitated calcium carbonate for paper filler applications. BioResources, 7(4), 5894–5903.

6 The Role of the Circular Economy in Mitigating Environmental Impacts of the Paper Industry

Saakshy

6.1 INTRODUCTION

According to the European Parliament, "the circular economy is a model of production and consumption which involves sharing, leasing, reusing, repairing, refurbishing and recycling of existing materials and products as long as possible". The Forests Stewardship Council says: "A circular economy is a systematic approach to economic development designed to benefit business, society and the environment" [1] (www.fsc.org).

Initially the concept of the circular economy started with three Rs: reduce, recycle, and reuse, as given in Figure 6.1.

Over time, the three Rs of the circular economy encompassed more framework and became the eight Rs of the circular economy. The eight Rs of the circular economy are given in Figure 6.2.

The transition from eight Rs to ten Rs of the circular economy is a transition on the basis of studies conducted by Kircherr and Potting. The ten principles of the circular economy are refuse, rethink, reduce, reuse, repair, refurbish, remanufacture, repurpose, recycle, and recover. The ten Rs of the circular economy are given in Figure 6.3.

6.2 PRINCIPLES OF THE CIRCULAR ECONOMY

The key principles of a circular economy include elimination or reduction of waste and pollution, circulation of products and materials, and regeneration of nature, as shown in Figure 6.4.

The circular economy aims to design out waste and pollution by rethinking the supply chain and challenging businesses to avoid waste and pollution. This principle focuses on reducing the environmental impact of production and consumption processes as well as keeping products and materials in use at their highest value for as long as possible. It involves processes such as maintenance, reuse, refurbishment, remanufacture, recycling, and composting. The circular economy also aims to

DOI: 10.1201/9781003461395-6

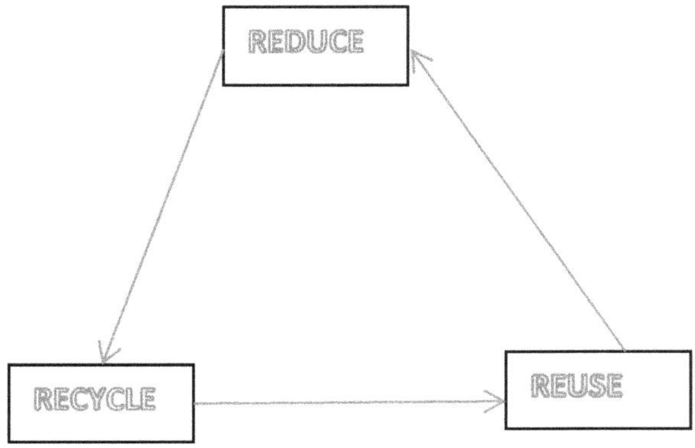

FIGURE 6.1 Three Rs of the circular economy.

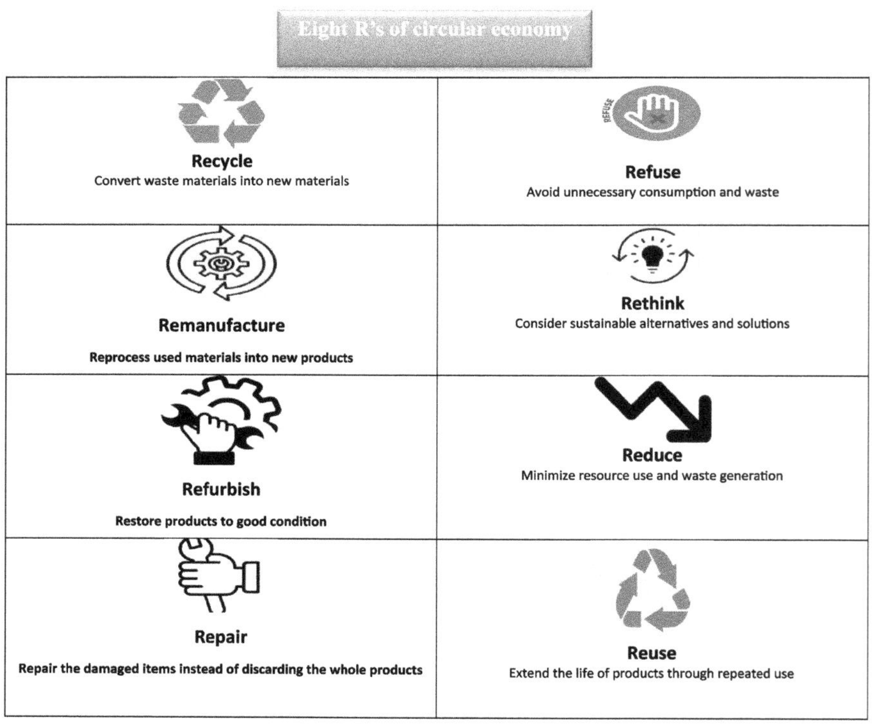

FIGURE 6.2 Eight Rs of the circular economy.

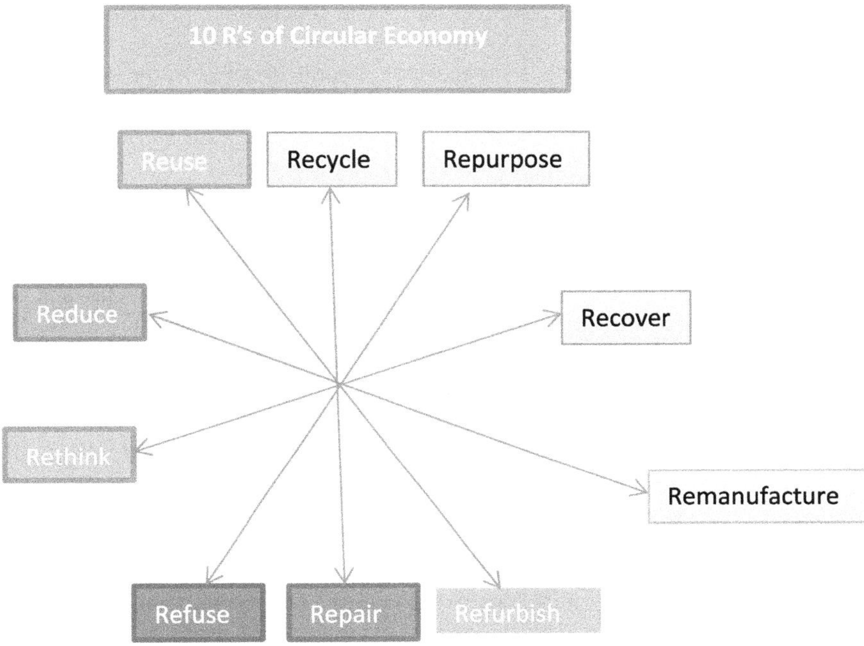

FIGURE 6.3 Ten Rs of the circular economy.

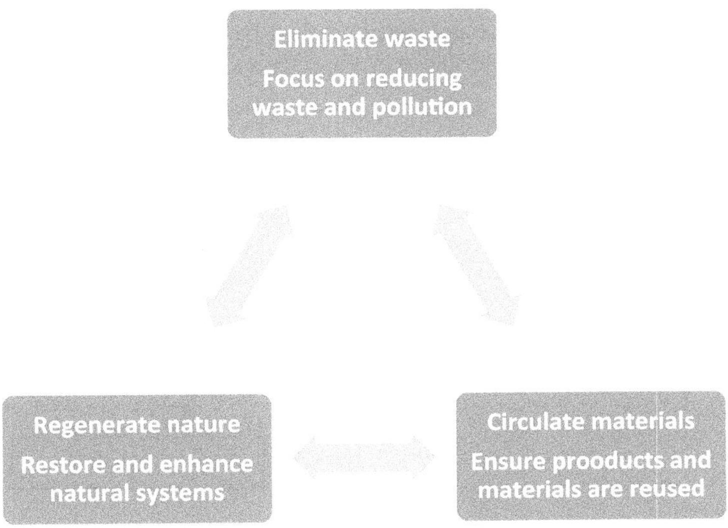

FIGURE 6.4 Principles of the circular economy.

minimize extraction of natural resources and waste generation by increasing the life of final products [2].

The circular economy seeks to regenerate natural systems by promoting sustainable production and consumption of goods. It advocates for the preservation or enhancement of renewable resources to support regeneration and utilization of renewable energy instead of completely relying on fossil fuels.

These principles underpin the transition to a circular economy, which aims to decouple economic activity from the consumption of finite resources and create a more sustainable and regenerative economic system. The circular economy model is designed to address global challenges such as climate change, biodiversity loss, waste, and pollution by promoting a more efficient and environment-friendly approach to production and consumption. The principles of the circular economy also include preservation, optimum consumption, and reduction of natural resources, achieving eco-design, development of local economy, putting an end to obsolescence, reducing the impact of human activity on the environment and health, and limiting consumption.

6.3 LIFE CYCLE OF PAPER

The first stage of the paper life cycle involves the cutting down of trees, which are then transported to a paper mill. The wood is then chopped into small pieces and transformed into wood pulp with the help of chemicals and water in a pressurized digester. The pulp is then washed to remove impurities and other unwanted materials. The washed pulp is then bleached to achieve the desired brightness, and finally, the pulp is converted into sheets in the papermaking section. The life cycle of paper is given in Figure 6.5.

Once the paper has been used for its intended purpose, it can be recycled. The used paper is collected, sorted, and cleaned before being processed into new paper. Recycled paper can go through the recycling process up to 6–7 times before the

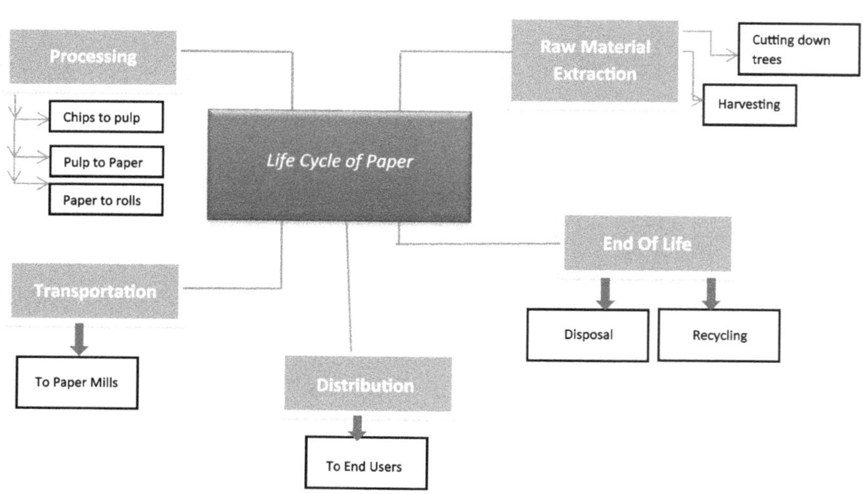

FIGURE 6.5 Paper life cycle: stages and processes.

cellulose fibers become incapable of forming sheets. The recycling of paper is an important part of the circular economy, as it helps to reduce waste and minimize the environmental impact of paper production.

6.4 ENVIRONMENTAL ASPECTS IN PAPER MANUFACTURING

The environmental impacts of paper production are significant, including deforestation by cutting of trees, air and water pollution, greenhouse gas emissions, and solid waste. Deforestation leads to habitat loss, species extinction, and climate change. The manufacturing process of paper releases nitrogen oxide, sulfur oxide, and carbon dioxide into the air, contributing to pollution such as acid rain [3].

Additionally, paper mills discharge various pollutants into surrounding water bodies, causing damage to aquatic ecosystems and threatening the health of people living near the mills. The production, transportation, and disposal of paper contribute to greenhouse gas emissions. When paper rots, it emits methane, a potent greenhouse gas, while combustion or composting releases carbon dioxide. Discarded paper and paperboard make up roughly 26% of solid municipal waste in landfill sites. According to the Nature Conservancy, the global average carbon footprint is close to 4 tons per person [4], and it is 4.7 tons of CO_2 per person according to the International Energy Agency (IEA) [5] and 4.8 tons per person in 2017 [6]. It is important to note that there are significant disparities in per capita emissions across different countries. For instance, the United States has one of the highest average carbon footprints, with an average of 16 tons per person, while India's average carbon footprint is approximately 1.89 tons per person [7] (www.worldometers.info, 2022). These variations are influenced by factors such as income, lifestyle, consumption patterns, and energy use. In summary, the average carbon footprint per person in the world is estimated to be around 4 tons of CO_2 emissions [8, 9].

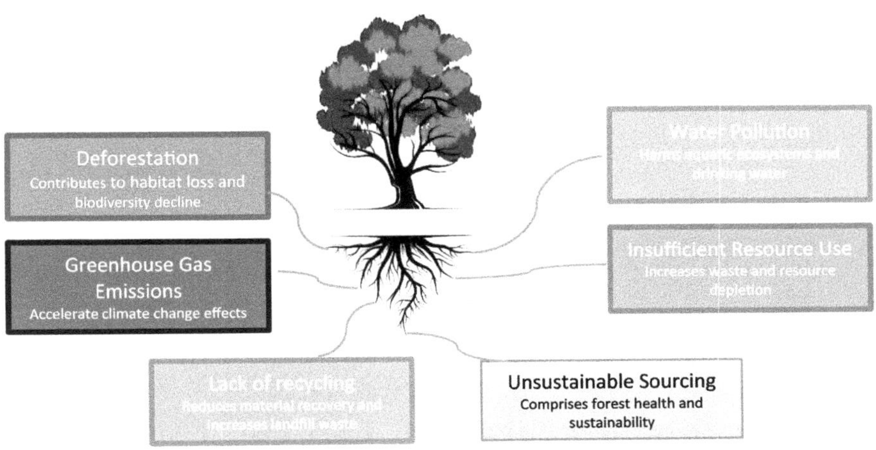

FIGURE 6.6 Environmental impact of the paper industry.

6.5 LINEAR ECONOMY OF THE PAPER INDUSTRY

The linear economy model of paper, as given in Figure 6.7, is a traditional model with the approach to the production, use, and disposal of paper. With time, this linear model has risked the sustainability, resource recovery, and environmental impact of the paper industry.

6.6 CIRCULAR ECONOMY IN THE PAPER INDUSTRY

The circular economy in the pulp and paper industry includes utilization of waste streams, maximization of resources, valorization of industrial waste, and transformation of the paper industry. The pulp and paper industry is transforming its operations to align with circular economy principles. This includes the industry's commitment to adopting circular economy practices, such as waste reduction and reuse, resource optimization, increasing the use of recycled fiber to replace virgin fiber, seeking innovative ways to improve resource efficiency, and valorization of industry wastes to create a more sustainable and regenerative economic system [3].

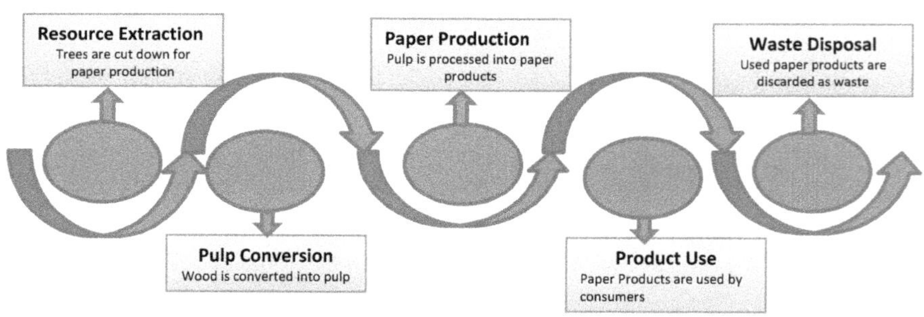

FIGURE 6.7 Linear economy model of paper industry.

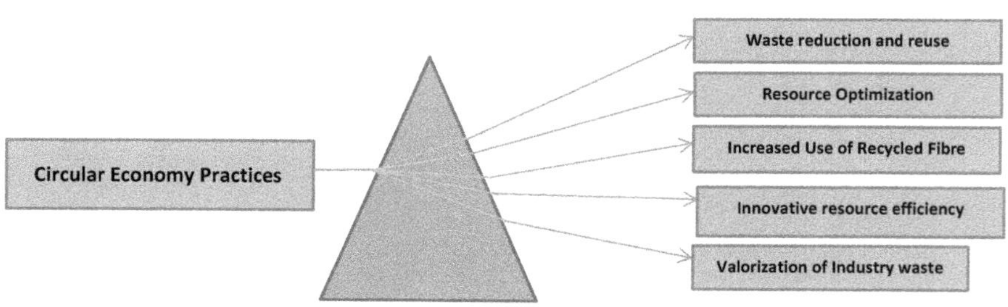

FIGURE 6.8 Circular economy practices for the paper industry.

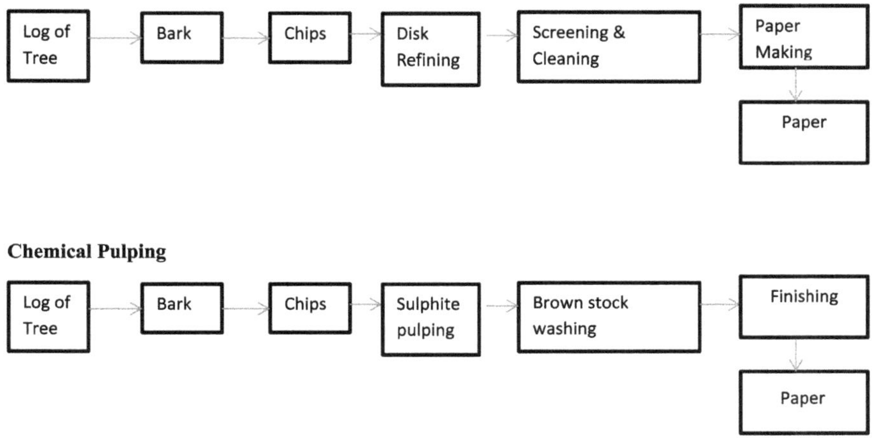

FIGURE 6.9 (a) Mechanical pulping; (b) chemical pulping.

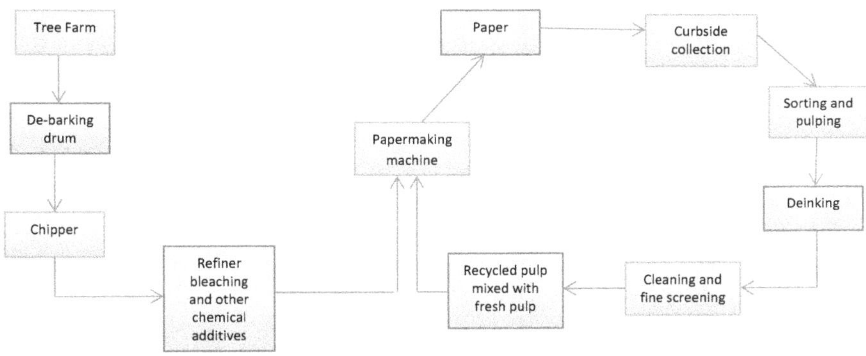

FIGURE 6.10 Pulp and paper manufacturing process.

Source: forestsbiofacts.com

6.6.1 Waste Reduction and Reuse

About 400 million tons of paper and paperboard are produced globally, and estimates suggest that global paper consumption in 2025 will amount to 500 million tons.

The various waste generated from the paper industry is given in Table 6.1.

The waste generated from the paper industry includes sludge, ash from boilers, wood waste, and wastewater with high biochemical oxygen demand and chemical oxygen demand and gaseous emissions.

TABLE 6.1
Waste Generated at Various Stages of the Papermaking Process

Main Stages	Processes	Waste Generated
Raw material	Handling and processing	Sawdust Bark Wood residues
Pulping	Digester	Black liquor, chemical waste
	Brown stock washer	Wash liquor
Cleaning	Primary, secondary, and tertiary centricleaners	
	Causticizing section	Lime sludge
Wastewater emissions	Bleaching	White liquor
	Pulping	Black liquor
	Washing	Wash liquor
Paper machine	Papermaking	Off cuts Trimmings
Wastewater treatment	Solid waste	Primary and secondary sludge 4–5% sludge 2.8% primary sludge and 1.2% secondary sludge
Air emissions	Gases from digesters, evaporators, and recovery	Nitrogen oxide (NO_x), sulfur oxides (SO_x), carbon dioxide (CO_2)
	Gases from boilers	Carbon monoxide (CO)
	Other air pollutants	Hydrogen sulfide, sulfur dioxide (SO_2), chlorine, and chlorine dioxide (ClO_2)
Boilers		Ash

6.6.2 Resource Optimization

Resource optimization is the more efficient utilization of raw material by optimizing production processes and reducing waste. The paper industry is highly dependent on raw materials such as wood, bamboo, and agricultural residues (wheat straw, rice straw, and bagasse). The shortage of raw material adversely affects the sustainability of the paper industry, including increases in production costs, logistics, chemicals, and raw materials. Due to the pandemic and the war in Ukraine, the entire paper business community has become mindful of the supply chain scenario for 2023. The issue of a conspicuous shortage of basic raw materials together with their high cost has aggravated the pressure on the margins and the viability of operations. There is an emphasis on increasing the recycling of paper and paper waste as a raw material for making paper and paper products.

6.6.3 Recycling and Reuse

The process of waste management in the paper industry begins with direct collection from various source points and shops where primary sorting of paper

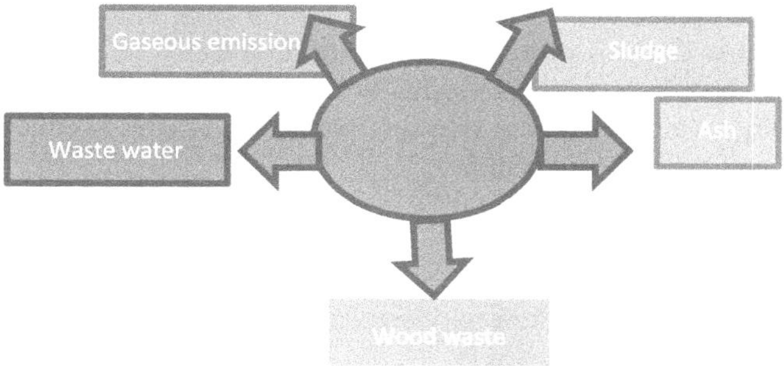

FIGURE 6.11 Waste generated from the paper industry.

waste into different grades of paper occurs. Then the paper waste is transported to zonal segregation centers, where the paper waste material is collected from shops and baled for dispatch to the end users. Indian waste paper collection is mostly carried out by the informal sector, rag pickers and door-to-door collectors. The concept of recycling has gained significant attention and momentum in recent years. The recycling of paper products and the use of recycled fibers in new paper production should be inculcated. Consumer awareness and education about the importance of recycling and efficient collection systems should be established.

6.6.4 Innovative Resource Efficiency

Incentivizing Recycled Content: Provides incentives for manufacturers to use recycled materials in their products. There are several countries, including the United States and Canada, that give incentives in the form of extended producer responsibility (EPR) schemes and regulations.

Developing High-Quality Recycled Products: During the process of paper manufacturing, investing in technology that has techno-economic feasibility in the recycling of paper and does not pose any environmental or health hazards, making it a viable alternative to virgin paper, should be emphasized.

Encouraging the use of recycled paper in production can reduce the demand for virgin fibers.

6.6.5 Sustainable Sourcing

The sourcing of raw materials from sustainably managed forests ensures responsible supply chain practices. The promotion of sustainable forestry practices is crucial for ensuring a continuous supply of raw materials without harming ecosystems. This includes following:

Certification Schemes: Supporting certification programs like the Forest Stewardship Council (FSC) that promote responsible forest management. In India, there is the Indian Forest & Wood Certification Scheme (IFWSC).
Agroforestry: Integrating trees into agricultural landscapes to enhance biodiversity and improve soil health.

6.6.7 Product Design

Designing paper products for longevity and recyclability makes it easier to reclaim materials at the end of their life cycle.

The principles and practices involved in creating paper products not only serve their purpose effectively but also facilitate material reclamation at the end of their life cycle. By focusing on innovative design, material selection, and production processes, we can contribute to a circular economy that minimizes waste and maximizes resource efficiency.

Material selection: Eco-friendly and recyclable material for making products should be selected. Recycled fibers and certified sustainable forest material should be used.
Production Process: The dyes and chemicals should not be harmful.
Innovative Design: Products should be developed which align with the recycling process as well as minimizing the use of polythene.

6.6.8 Innovative Business Models

Adopting innovative business models can help shift the paper industry towards a circular economy. Examples include:

- **Product-as-a-Service**: Offering paper products as a service rather than a one-time purchase, encouraging reuse and recycling.

FIGURE 6.12 Sustainable paper product design.

- **Take-Back Programs**: Implementing take-back schemes where consumers can return used paper products for recycling or repurposing.

6.6.9 Valorization of Paper Industry Wastes

Valorization in the paper industry includes the conversion of paper industry waste to valuable resources like fuels, chemicals, or value-added products.

6.6.9.1 Fiber Recovery

The fiber losses from the paper industry can be recovered and used as insulation and construction materials.

6.6.9.2 Composting

Paper waste sludge can be used as potential nutrient-rich compost. This improves soil health and reduces the need for chemical fertilizers.

6.6.9.3 Recovery of Chemicals

The recovery of chemicals used in the papermaking process can be achieved through various methods allowing their reuse in production, thereby reducing production cost.

6.7 Methodology and Results: Multi-Criteria Analysis

Multi-criteria analysis is a decision-making tool that helps in evaluation of multiple criteria or factors. Multi-criteria analysis helps in the assessment of the performance of different strategies.

6.8 Conclusion

Circular economy 3.0 emphasizes economic growth and environmental stewardship. The circular economy emphasizes an increase in the utilization of recycled fibers,

TABLE 6.2
Multi-Criteria Analysis

Criteria	Parameters	Factors
Quantitative	Environmental	Greenhouse gas emissions Wastewater Solid waste
	Economic	Capital cost
Qualitative	Societal	Garbage
	Technological	Energy consumption
	Regulatory compliance	Pollution parameters

TABLE 6.3
Methodology of Multi-Criteria Analysis

Step 1	Define objectives	Clearly outline the objective goals of analysis considering the circular economy
Step 2	Identify criteria	Select relevant criteria that will be used to access the options available for circular economy initiatives
Step 3	Weight criteria	Assign weights to each criterion based on their importance to the stakeholders involved
Step 4	Evaluate options	Analyze different circular economy strategies
Step 5	Aggregate results	Combine evaluation to get overall scores
Step 6	Sensitivity analysis	Find the effects of the evaluation and criterion weights

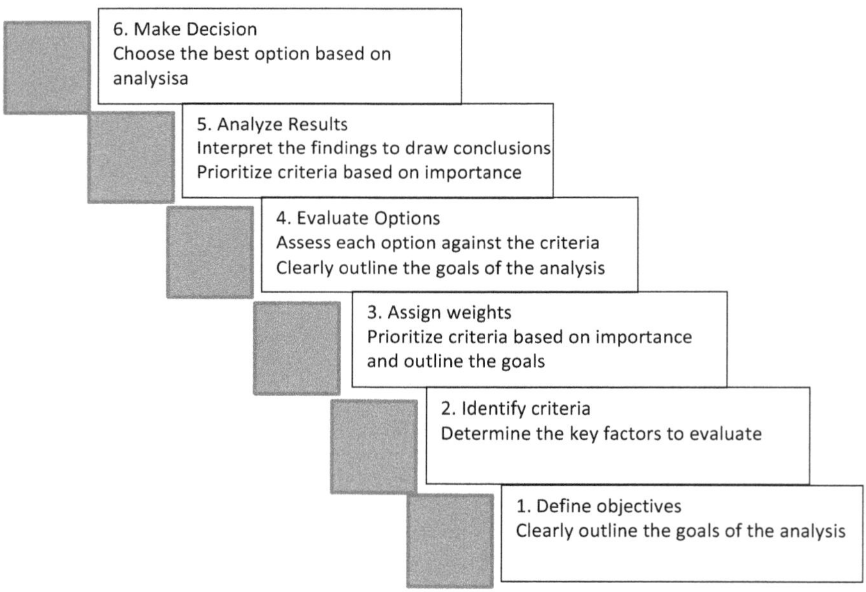

FIGURE 6.13 Stepwise methodology of multi-criteria analysis.

thus reducing the impact on forests and eco-systems. It is pertinent in the present scenario to reduce environmental footprints through resource efficiency, recycling, and technological interventions to make new products like construction materials, nano-cellulose, adhesives, cementing materials, antibacterial agents, phenolic compounds, composite materials, bioethanol, hydrogen and biofuels, and cellulose nano-fibers that will create new marketing avenues and reduce carbon footprints.

REFERENCES

[1] www.fsc.org

[2] The Impossibilities of Circular Economy: Separating aspirations from reality. (2022). Ed. Harry Lehman, Christoph Hinske, Victoire de Margerie, Anetaslaveikova Nikolova, Routledge, 1st Edition, 1–332. ISBN-9781003244196.

[3] Jiang, S., Li, B., & Shen, Y. (2021). The influence of pulp and paper industry on environment. *Environmental Ecology & Biochemical Testing*, E3S web of conferences 308, 6th International conference on Materials Science, Energy Technology and Environmental Engineering (MSETEE), 02007.

[4] Sun, Y.-Y. (2023). Carbon footprint factsheet. Center for Sustainable Systems, University of Michigan, Pub No. CS, 411–414. https://doi.org/10.4337/9781800377486.carbon.footprint

[5] IEA. (2023). *CO_2 Emissions in 2022*, OECD Publishing, Paris. https://doi.org/10.1787/12ad1e1a-en.

[6] Hannah, R. (2019). Where in the world do people emit the most CO_2? *Our World in Data*, 1–5. https://ourworldindata.org/per-capita-co2

[7] www.worldometers.info

[8] Cozzi, L., Chen, O., & Hyeji, K. (2021). The world's top 1% of emitters produce over 1000 times more CO_2 than the bottom 1%—Analysis—IEA. 3–10. https://www.iea.org/commentaries/the-world-s-top-1-of-emitters-produce-over-1000-times-more-co2-than-the-bottom-1

[9] Friedlingstein, P., Sullivan, M. O., Jones, M. W., Andrew, R. M., Bakker, D. C. E., Hauck, J., Landschützer, P., Quéré, C. L., & Luijkx, I. T. (2023). Global carbon budget 2023. ESSD 15(12), 5301–5369.

7 Case Study
Preparation of Fly Ash–Based Filler in Papermaking

Suresh Sundaramurthy and Anuj Kumar Verma

7.1 INTRODUCTION

Most writing printing and other paper grade products contain at least two major components. The first is fibers, and the other is filler or pigment (Hubbe & Gill 2016). The main function of fiber is to provide a fiber network. This fiber network is usually held together by inter-fiber hydroxyl bonds, thus providing intrinsic strength (Alava and Niskanen 2006), whereas the main function of filler is to import aesthetic properties such as brightness, opacity, and tear index, and yet another function is to fill the space between the pulp fibers and thus help to produce a smooth and printable surface (Shen et al. 2009). However, increasing the filler content in any paper grade decreases the inter-fiber bonding, and wet end parameters such as FPAR and FPR must be taken into consideration when different kinds of filler are used in research (Chen et al. 2011). Production with high brightness novel calcium silicate made from fly ash are gaining more attention in papermaking as a filler. This works reports the utilization of calcium silicate filler and pigment made from fly ash in writing printing paper and newsprint.

7.1.1 Papermaking

Wood, rags, or grasses and agricultural residue are the basic raw material for papermaking. In developing countries, 60% of the raw material comes from non-wood raw material such as bagasse, cereal straw, bamboo, reeds, esparto grass, jute, flax, and sisal (Gullichsen, 2000). It is made by the dilute aqueous suspension of fiber through a mat of wire and felts. Last, this formed sheet pressed and dried. In papermaking, there are two major components, fiber and filler (Solomon and Hawthorne, 1983). In recent years, the demand for high-quality paper at low prices is increasing as digital media is putting lots of pressure on various paper grades like newsprint and writing printing paper, so filler engineering has become an important part of current research.

7.1.2 Newsprint, Writing Printing Paper, and Paperboard Production in India and the World

Only 125–130 Newsprint papermills were controlled by the Newsprint Control Order 2004. At present, per the reported data, 3.3 million tons of newsprint are produced

DOI: 10.1201/9781003461395-7

per annum from 64 mills; however, 23 mills are closed, and 36 mills have stopped newsprint production (Mathur et al. 2009). Thus, there was a sharp drop in domestic newsprint production from 1.44 million tons to 1.02 million tons from the years 2014–2015 to 2015–2016. Therefore, India is importing half of its newsprint from other countries, and these imports significantly increased from 1.33 to 1.5 million tons from the years 2014–2015 to 2015–2016 (Jain et al. 2017). There is increasing demand for improved newsprint properties because advertising is switching from more expensive paper grades to newsprint without losing quality. The export of newsprint from India is negligible, so there is an urgent need for cheap pigment as a coating material. In addition to that, the industry is facing tremendous financial pressure due to lower-cost imports. The writing printing paper industry's production has shrunk from 15 to 10 million tons per year in the last 5 years (DGCI&S, 2016) According to the 2024–2025 report from the Directorate General of Commercial Intelligence and Statistics (DGCI&S), India is importing more than it is producing and tabulated in Table 7.1.

Natural minerals are the basic raw material to produce paper filler and coating pigment. Carbon and calcium are the main elements present in natural minerals. The particle sizes of various fillers used in paper industry are in the range of 2 to 10 μm. Larger and rougher materials are considered a filler material, while smaller and softer materials are considered pigments. Each filler has its own advantages and disadvantages depending upon the end use. Sometimes the choice of filler in papermaking may be the combination of two fillers. A comparison of the cost and performance of various types of filler was done by Mueller (2005).

TABLE 7.1
Import and Export Pattern (DGCI&S)

India's Imports of Paper & Paperboard (DGCI&S)

	Thousand Tonnes			
Year	**Total**	**China**	**ASEAN**	**S Korea**
2020-21	1.084.7	145.9	145.2	63.8
2021-22	1,145.8	145.8	148.3	64.7
2022-23	1,436.5	341.0	266.5	81.7
2023-24	1,929.1	411.4	514.1	103.4
2024-25	2,055.3	547.3	415.1	99.4
4 Year CAGR (%)	17.33	39.17	30.03	11.72
2024-25 (April to June)	451.8	111.6	77.8	21.2
2025-26 (April to June)	486.3	142.6	91.5	18.4
Growth over Last Year (%)	7.64	27.78	17.61	–13.21

7.2 MATERIALS AND METHODS

The raw material used for newsprint sheets was a blend of 97% old newsprint (ONP), 2% old magazines (OMG), and 1% coated book stock (CBS). The raw material used for writing printing sheets was a blend of 25% broke, 15% bleached softwood pulp, and 60% bleached hardwood pulp in some cases. The fly ash used in this study was taken from a coal-fired power (thermal power) plant in Jhansi, Uttar Pradesh, India.

7.2.1 Coagulation (Charge Neutralization)

Charge neutralization is achieved with the help of coagulation. In papermaking, if lots of negative charges are developed due to fines and a closed-loop system, then positive charge coagulation is used for charge neutralization and vice versa (Gallagher, 1990). Polyvalent cations, polyethyleneimine (PEI), polyamines and polyamideamine epichlorohydrine (PAE), and polydiallyldimethylammonium chloride (poly-DADMAC) are most frequently used for charge neutralization (Norell et al., 1999; Cadotte et al., 2007). In the present study, N-7607 low molecular weight cationic coagulant is used.

7.2.2 Flocculation (Bridge Mechanism)

After coagulation, that is, for charge neutralization, flocculation plays an important role to patch up particles. This is achieved by a bridging mechanism (Gregory, 1985; Berlin and Kislenko, 1995; Biggs et al., 2000). The bridging mechanism depends upon the flocculent molecular weight, charge density, and concentration (Berlin and Kislenko, 1995; Berlin et al., 1997; Bremmel et al., 1998; Blanco et al., 2002). A combination of coagulation and flocculation helps in superior retention and drainage (Eklund and Lindström, 1991; Cadotte et al., 2007). In the present study, N-7530 cationic flocculants are used.

7.2.3 Microparticle Retention Aid Other Additive

In high shear paper machines, coagulation and flocculation are not sufficient for good retention and drainage, as the flocculation bridge mechanism breaks down due to high shear conditions. In order to overcome this difficult problem, the majority of papermakers use microparticles to re-flocculate the pulp furnish so that shear conditions do not affect retention and drainage performance (Clémencon & Geril, 1999). Microparticle systems have several advantages over dual polymer system (coagulation plus flocculation), such as fast dewatering and high porosity for the final paper (Wågberg et al. 1996). The most common microparticle used in the papermaking process is colloidal silica and microparticle Bentonite (Na-Bentonite) (Swerin et al. 1993). In the present study, microparticles (Bentofloc) was supplied by an industry in the northern part of India.

Starch is the most important additive used in papermaking due to its hydrogen bonding between cellulose hydroxyl groups and starch molecules, which results in improved strength and surface properties of the final product. Starch does not affect the optical and surface properties of the paper, but the uniformity of the paper may be disturbed by excessive starch. In the present study, a high degree of substitution (DS = 0.84) starch supplied by a Thailand-based company is used.

7.2.4 Surface Sizing Agents

In order to save paper from water as well as ink, a surface sizing agent is used in the paper manufacturing process to provide water resistance. The efficiency of the sizing chemical depends upon the sizing chemical retention with fines (when no filler is used), and with filler, its efficiency depends upon both fine and filler retention (Lindström and Söderberg 1986b, Isogai 1992, Champ and Ettl 2004, Johansson and Lindström 2004a, b; Ravnjak et al. 2007). The most common internal sizing agents used in the paper industry are alkyl ketene dimer (AKD) and alkenyl succinic anhydride (ASA). AKD is a mixture of stearic (18 C-atoms) and palmitic (16 C-atoms) acids, whereas ASA is synthesized from olefins with a chain of 16–18 carbon atoms.

The sizing depends on the following:

1. Redistribution and spreading of sizing chemical on filler and fiber surfaces
2. Act as an anchor between filler and fiber surfaces
3. Retention of size particles of filler and fiber

The reaction of ASA with water is as follows:

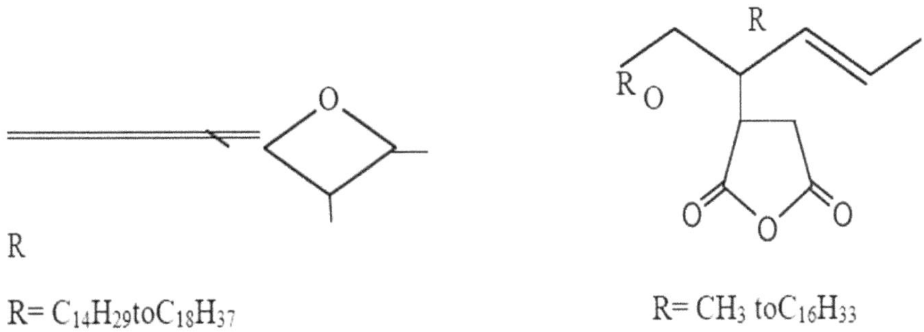

FIGURE 7.1 Molecular structure of AKD and ASA.

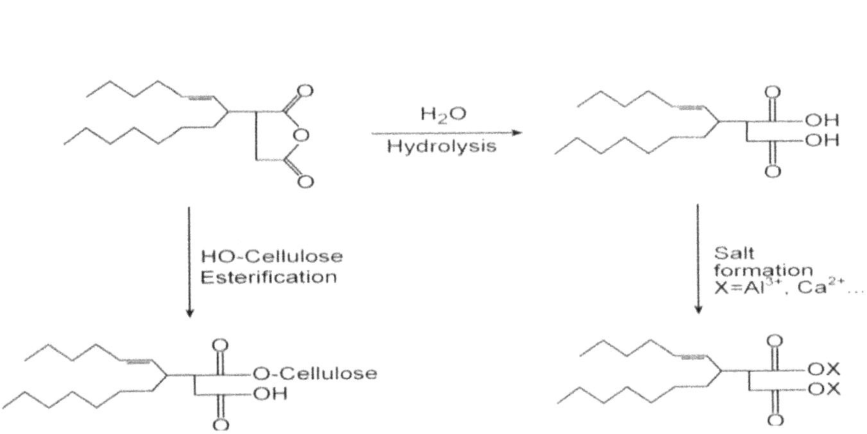

FIGURE 7.2 Main reaction of AKD is with cellulose, water, and itself to form oligomers.

Filler engineering has many advantages; however, when high-ash paper is produced, paper strength properties are reduced due to the high filler percentage. In order to overcome this limitation, filler treatment before the addition to pulp is an important area of study.

7.3 VARIOUS ROUTES OF FILLER TREATMENT

In recent years, high filler content has become important in the pulp and paper industry to reduce overall production costs. The addition of filler can improve paper optical properties, sheet formation, paper dimensional stability, printability, and appearance (Yuan 2008; Shen et al. 2007; Lu et al. 2010). However, filler increases have several disadvantages and limitations (Han 2009; Shen et al. 2009) such as fiber–fiber bonding, low retention, high RDF dosages, and low bulk.

Although traditional filler technology has both advantages and disadvantages, filler modification using starch, cellulose derivatives, chitin or chitosan, guar gum, and many other specialty chemicals has become an active area of research to overcome these difficulties. Many approaches has been used or studied previously to increase the proportion of filler without sacrificing paper properties such as pre-flocculation of filler, modification of filler, lumen loading, synthesized filler, and using specialty polymer in the sheets (Li et al. 2016).

1. Filler treatment with starch
2. Filler treatment with phosphoric acid
3. Filler treatment with anionic trash controller (ATC)

7.3.1 Filler Treatment With Starch

Starch has good bonding ability with wood fibers due to its three hydroxyl groups on each glucose monomer unit (Deng et al. 2008). Starch can be dissolved and either adsorbed or precipitated by appropriate methods. During the adsorption or precipitation, this bonding material can be located on the surface of the filler, by which the paper strength will be improved by hydrogen bonding between the fibers and fillers (Yoon 2007; Zhao 2008; Yan 2005). Shen et al. (2008) modified PCC filler using a complex formation method with raw starch and aluminum sulfate. Huiming et al. (2012) modified filler using a starch–sodium stearate complex and also characterized its particle size with a particle size analyzer and its morphology by SEM.

7.3.2 Filler Treatment With Phosphoric Acid

Phosphoric acid is extensively used as a modifier by researchers (Peng et al. 1998; 2001; 2003). Phosphoric acid can be used as a modifier for a filler to make it suitable to blend with fiber without destroying any paper properties. Phosphoric acid has some advantages and disadvantages when used as a modifier. The major advantages and disadvantages are:

1. The acid-resistant properties of calcium carbonate filler can be enhanced when it is treated with weak acid such as phosphoric acid due to the pH buffering effect (major advantage).

2. Phosphoric acid–modified filler also increases the overall brightness of the paper.
3. Scale formation of calcium phosphate (disadvantage).
4. Release of unwanted phosphorous to wastewater (disadvantage).

Preferred dosages of phosphoric acid were found to be 0.2 wt% based on weight of $CaCO_3$ (Pang, 2001).

7.3.3 Filler Treatment With Anionic Trash Controller

The anionic trash catcher gathers electrolytes for strong cations in the paper-making process. The anionic trash catcher can be used as an anion capture agent and can effectively increase the size of the effect of filler to increase the retention rate and solve many anion problems. The anionic trash controller can also improve the performance of paper machines, extend the life of felts and wires, reduce steam consumption, and improve speed and production.

7.3.4 Uses of ATC

The anionic trash catcher can be used in copy paper, newsprint, coated paper, kraft paper, and other paper. The situation is different when adjusting zeta potential to the range of 2–6 mV through polyelectrolyte charge detection (PCD); zeta potential and concentration should be adjusted appropriately according to the specific technology. The anionic trash catcher can be used with starch, polyacrylamide (cPAM), and other retention and drainage aids, and can be added directly to the pulp or white-water system.

7.3.5 Grinding and Fractionation of Fly Ash Based on Particle Size

The fly ash sample was ground in a ball mill to achieve a uniform fine size, and then it was passed through stacks of sieves with mesh sizes of 200, 250, 300, 325, 400, 500, and 800. The sieving operation had a duration of 10 min, and after the shaking was completed, the material on each sieve was weighed. The particles in the size range from 21 to 39 μm were used as the pigment and fillers for papermaking.

7.3.6 Unwanted Material Separation From Fly Ash for Whiteness Improvement

Magnetic pearls, unburned carbon, and mullite content are the major influencers on the whiteness of fly ash. Magnetic pearls are formed during the decomposition of coal minerals, such as magnetite arsenic, pyrites, and magnetite. During combustion, Fe_3O_4 and magnetic Fe_2O_3 form magnetic pearls, which are black in color and have strong magnetic properties. The average sizes of these pearls are 18 to 22 μm, with a whiteness of only 9.1. These pearls present in fly ash have a high influence on its whiteness. The non-magnetic fraction of fly ash was separated using magnets.

Another component that affects fly ash whiteness is unburned carbon particles. The content of unburned carbon particles in the present study was only 2.61%. Because the ultrafine unburned carbon particles are grey and black in color, they have a large effect on the whiteness of fly ash. A self-designed floatation separator was used to remove this unwanted material from the fly ash. The success of the flotation process depends on the collector dosages. In this experiment, the optimum collector dosages of 15 ml of kerosene per 1.5 kg of fly ash were used. The fly ash whiteness was measured using L&W Elrepho Code Number 070/071 (Kista, Sweden).

7.4 STEPS INVOLVED IN FILLER AND PIGMENT MANUFACTURING

7.4.1 Step 1: Digestion

This step involved the digestion of fly ash with sodium hydroxide; here, the caustic acted as an activator in specific conditions. In this process step, silica was extracted from fly ash with sodium hydroxide to form two products: a soluble sodium silicate solution and solid aluminum oxide.

$$\text{Fly ash} + \text{NaOH (1)} \rightarrow \text{Digested ash} + \text{Undigested ash}$$

Here, the digestive ash was a soluble sodium silicate solution, whereas the undigested ash was converted into aluminum oxide. After completion of the step, the solution was separated out, and this separated clear filtrate was taken for the precipitation step.

Fly ash was treated with an activated agent like sodium hydroxide for 1.5 hours at 90°C. This makes the silica present in fly ash soluble in caustic solution, while the undigested ash remains as a solid product consisting primarily of aluminum oxide. The percentage of silica extract from fly ash depends upon the concentration of the caustic. A series of experiments were performed and found the optimal conditions for the digestion of fly ash with a caustic.

In the NaOH leaching method, 20 grams of NaOH were dissolved in 600 ml of deionized water, and then 20 grams of fly ash were added to the solution. These mixtures were stirred and maintained a temperature of 90°C for 1.5 hours. For optimization of the experiment, we took two routes:

1. Fly ash weight, time, and temperature were kept constant, while the NaOH concentration varied from 0.83 to 12.5 M.
2. The fly ash mass varied from 10 to 40 grams, while the NaOH concentration of 7.34 M, deionized water 600 ml, time 90 minutes, and temperature 90°C were kept constant.

7.4.2 Step 2: Precipitation

In this step, silica was extracted from the sodium silicate solution. Carbon dioxide (99.98% pure) was passed through the sodium silicate solution under different operating conditions. During the reaction, continuous stirring was applied in order to get the precipitated silica.

$$Na_2xSiO_2(l) + CO_2 \longrightarrow xSiO_2(s) + Na_2CO_3(l)$$

After step 1, the sodium silicate solution was separated out and kept in a high-pressure vessel, in which CO_2 gas was passed out at 95°C until the pH reached 8.5. After stopping the carbonation, sodium carbonate was filtered out, whereas the precipitated silica was washed to remove soluble salt (98.9% purity silica was achieved) and dried for making filler.

7.4.3 Step 3: Hydrothermal Reaction

This reaction was carried out in a high pressure vessel and is also known as a hydro thermal reaction. In this process, we generally observe three steps. First, silica and lime slurry are formed under different operating conditions separately. Second, the concentration of two slurries $(CaO \,\&\, SiO_2)$ is adjusted in the final mixture to get the desired product. Third is stopping the reaction and cooling down.

I. Silica Slurry formation

$$(SiO_2)_y + 2y\ H_2O \longrightarrow y\ Si\ (OH)_4$$

II. Lime Slurry Formation

$$CaO + H_2O \longrightarrow Ca\ (OH)_2$$

Combine reaction (hydrothermal reaction)

$$x\ Ca(OH)_2 + y\ Si\ (OH)_4 \longrightarrow (CaO)_x\ (SiO2)_y.(x+y)H_2O$$

The values of x and y play a vital role in producing a series of minerals formed from the combination of calcium and silica. Various forms of calcium silicates are tobermorite, foshagite, riversideite, and xonotlite.

The formation of these different minerals is dependent on several factors. The most important ones are the highest temperature and pressure of the reaction, the concentration of the solids in the slurry, and the ratio of calcium to silica. To make filler that reduces drying energy as well as fiber substitution, a bulk temperature of 190°C is needed in order to provide adequate energy to drive the reaction. The different temperature conditions are needed based on the different silicate products made.

The ratio of calcium atoms to silica atoms is key to the growth of the right crystal structure. Each of the mentioned minerals has a different ratio of calcium atoms to silica atoms. When the ratio is low, at about 0.85, riversideite is formed. When the ratio is about 1.00, tobermorite is made. Finally, if the ratio is high, at 1.30, then foshagite is grown. Each of these crystals has a different structure and affects the properties of paper in a different way.

7.5 FILLER CHARACTERIZATION

The particle size distribution of both fly ash–based calcium silicate fillers and pigment was measured using a laser scattering PSD analyzer (Malvern Mastersizer 2000). Filler brightness was measured using TAPPI method T 534 pm–92. A pH meter (Manti Lab MT-103 Digital pH meter) was used to measure the pH of fillers manufactured using the high-pressure carbonation process.

The specific surface area of fillers was measured by BET analyzer (SA 9600 series). Various properties of the modified fly ash–based calcium silicate as a pigment and marketed PCC are given in Table 7.2, while various properties of FACS used as filler and other marketed PCC used are shown in Table 7.3.

TABLE 7.2
Various Parameters of Calcium Silicate and PCC as a Pigment

Properties	FACS	PCC
Average particle size (μm)	15.8	1.75
BET area (m^2/gm)	83	11.2
Brightness (ISO)	91.2	96
Solid (%)	3.5	16
pH	8.1	9.2
Refractive index	1.64	1.57

TABLE 7.3
Characteristics of FACS and PCC Filler

Characteristics	FACS	PCC
Average particle size (μm)	4.8	1.9
Particle size distribution	1.12	1.11
Specific surface area (m^2/g)	151	14.3
Brightness (%ISO)	93.8	96.4
pH	8.3	9.1

7.6 COATING FORMULATION FOR ONP SIZE PRESS COATING AS A PIGMENT

The coating formulation was made using a high shear mixer with a variable drive to provide high shear for dispersion of the pigments but low shear for the addition of the latex and starch. The final coating formulation was then diluted with water to provide a coat range close to the target of 4.5 and 7.5 gsm on a coater with a Brookfield viscosity of approximately 900 cp for both FACS and PCC. After the sheets were coated, they were immediately dried using a coated drum dryer, calendared, and then placed in a standard control room for testing. The coating formulation for the ONP on the size press is given in Table 7.4A shows that the final coat weight for this experiment was 4.3 gsm and 7.7 gsm (±0.5 gsm). Four coating pigment formulations were prepared: FACS-4.3 gsm (F1), FACS-7.8 gsm (F2), PCC-4.2 gsm (P1), and PCC-7.6 gsm (P2), as shown in Table 7.4B.

TABLE 7.4A
Details of Coating Formulation

Constituent Parts	Parts
Fly ash–based calcium silicate pigment	100 parts
Latex	5 parts
Starch	7 parts
Total	112 parts

TABLE 7.4B
Newsprint Various Properties Before and After Coating (significant differences at p < 0.06)

Types of Paper	Brightness (ISO)	Opacity (ISO)	Gurley Stiffness (mg)	Tear Index ($mN.m^2/g$)	Breaking Length (km)	Smoothness (su)
Newsprint sheet as such	50.2 ± 1.2	98.1 ± 3.2	25.9 ± 4.2	5.1 ± 0.5±	5.2 ± 0.8	226 ± 12
F1	63.4 ± 2.3	99.2 ± 4.3	28.8 ± 3.2	5.9 ± 0.8	6.8 ± 0.1	248 ± 10
F2	73.2 ± 1.7	99.6 ± 2.1	35.8 ± 1.2	5.6 ± 0.3	6.9 ± 0.4	295 ± 16
P1	67.4 ± 2.1	99.5 ± 2.7	26.8 ± 3.4	5.3 ± 0.4	5.9 ± 0.6	284 ± 13
P2	79.8 ± 1.8	99.8 ± 3.0	29.4 ± 2.3	5.5 ± 0.2	6.3 ± 0.2	327 ± 18

7.7 FILLER TREATMENT AND PRE-FLOCCULATION FOR STRENGTH IMPROVEMENT

Various approaches have been investigated to address the challenge of increasing ash content in sheets. In the current study, the concept of pre-flocculation of fillers with the help of cationic starch, an anionic trash controller, and its particle size distribution using Malvern Mastersizer 2000 prior to their addition to the wet end has been analyzed. The term pre-flocculated means the modification of filler particles in agglomerates through treatment with starch before adding to fiber. This balanced modification is expected to lead to a reduced negative impact of the filler addition on paper strength. Due to high-speed paper machines, filler size and flocs should be stable and shear resistant (Cheng et al. 2011). The starch was heated at 3.0% solids for at 160°C (inlet temperature) and 40°C (outlet temperature), mixed with 10% solid fly ash–based calcium silicate slurry, and stirred at 220 rpm. The treated filler particle size and particle size distribution were checked by a Malvern Mastersizer 2000 instrument.

7.8 RETENTION AND DRAINAGE TEST

A Britt dynamic drainage jar (BDDJ) device with a standard steel screen (hole diameter = 76 μm) at 750 rpm was used to measure the first pass ash retention (FPAR) and first pass retention (FPR). FPR and FPAR is calculated as follows:

$$\text{FPR}(\%) = \frac{\text{Consistency of Stock} - \text{Consistency of DDJ Filtrate}}{\text{Consistency of Stock}} \times 100$$

$$\text{FPAR}(\%) = \frac{\text{Fillers in Stock }(\%) - \text{Fillers in DDJ Filtrate }(\%)}{\text{Fillers in Stock }(\%)} \times 100$$

7.9 RETENTION AND DRAINAGE

Retention is the final amount found in the end product relative to the amount of substance in the initial steps (Scott, 1996). In the paper industry, retention is measured by first pass retention and first pass ash retention (Norell, Johansson, & Persson, 1999). First pass retention, or wire retention, is measured in every 2 hours or varies from mill to mill.

The major purpose of using retention aids is to make filler size large to agglomerate with fiber particles and fines so that it will not drain with the water from the wire hole. It is always advisable to use a filler size greater than the wire holes in order to reduces losses (Seppänen, Ström, & Elftonson, 2000; Blanco, Negro, Fuente, & Tijero, 2005). The first retention aid used by papermakers was alum (Norell, Johansson, & Persson, 1999). In brief, the function of retention aids is to:

- Accelerate drainage, hence increasing machine speed
- Require less time for cleaning, giving overall high production
- High filler and fine retention
- High pulp brightness
- Clean closed water system
- Fewer deposit problems
- High machine runnability
- Less drying energy

7.9.1 Troubles With Low Retention

- More fines and contamination generation
- More additive dosages are required
- Decreased runnability of machine due to breaks
- Deposits in press felts
- Poor paper sheet formation
- Increase in effluent treatment plant (ETP) load

7.10 ANALYTICAL METHODS

7.10.1 Brightness and Whiteness

L&W Elrepho code 071 (Sweden) is a spectrophotometer specially designed for the paper industry. It measures opacity, brightness, color, and whiteness. A sample is placed in the measurement position, then the sample measurement is taken automatically and the results appear on the computer screen. These results are notes in a laboratory book. It is also useful in the measurement of the ERIC value of deinked or recycled pulp. Central laboratory equipment from JK Paper Songadh and Emami Paper Mills Balasore was used to measure optical properties.

7.10.2 Tear Resistance Tester

The tearing index is the ratio of tearing strength to the grammage of the paper sheet. Its unit is $mN \cdot m^2/g$.

$$\text{Tearing index} = \text{Tearing strength/grammage}$$

Tear strength depends on fiber length, fiber, and pigment bonding (Chauhan et al. 2013). Its depends upon the porous structure of the filler material when blended with fiber material; surface roughness also plays an important role in the tear index (Chauhan et al. 2013).

7.10.3 Particle Size Analyzer

It is based on the phenomena of laser diffraction (Mie theory). It measures the particle size in the range of nanometers to millimeters. It describes the size of the particle (spherical) in terms of maximum length, minimum length, sedimentation rate, sieve aperture, surface area, and volume. Laser diffraction-based results represent volume-based particle size distribution.

7.10.4 Burst Strength and Sizing Testers

It is a tool for measuring the strength of paper, paperboard, and paper boxes. The Cobb value represents the water absorption capacity. A lower Cobb value is preferential for highly waterproof paper sheets (Sharma et al. 2012). The Cobb value also depends upon the amount of hard and soft wood in papermaking (Biricik et al. 2011).

A high-pressure carbonation vessel is used for the chemical reaction.

7.10.5 Hand Sheet Preparation and Testing

A laboratory sheet former is used to manufacture 42 gsm (g/m^2) newsprint sheet and 70 gsm writing printing sheet for experimental purposes; here, filler was pretreated with starch at 5 to 15% ash content in a newsprint sheet, whereas filler

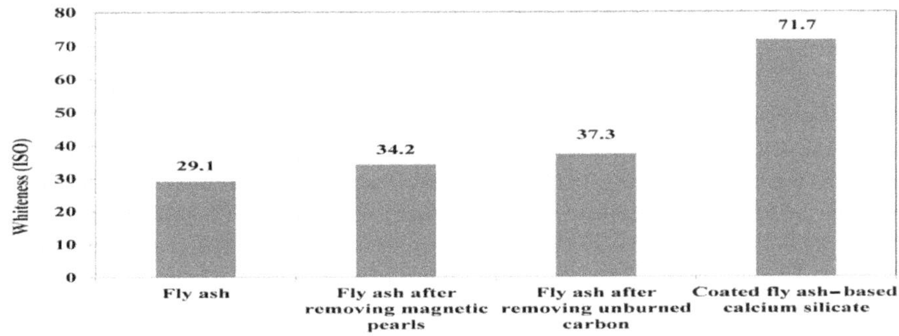

FIGURE 7.3 Whiteness gain after various process.

was treated with starch at 15 to 25% ash content in writing printing paper per TAPPI Test Method T-272 sp–97. Drying and pressing were done according to TAPPI Test Method 218 sp–02. The conditioning of sheets was done at relative humidity for 4 hours following TAPPI Test Method T402sp–98. Newsprint and writing printing paper ash content at different filler loadings ranged from 5 to 25%, measured using TAPPI Test Method T211 om-93 at 525°C. CPAM retention aid was added to the oven-dry pulp at a dosage of 0.05 wt%. The porosity of newsprint sheets was evaluated per TAPPI test Method T 460 om–96. The newsprint base sheet was conditioned for 3 h in an environmental chamber maintained at 25°C and 52% relative humidity. Brightness (TAPPI T 452 om-98 1998), opacity (TAPPI T425 om-96 1996), tear index (TAPPI T414 om-98 1998), and Gurley stiffness (TAPPI T543 om-00 2000) were calculated using the TAPPI standard methods.

The whiteness of unmodified fly ash was 29.1 ISO, as shown in Figure 7.3. After removing magnetic pearls and unburned carbon, the whiteness was 37.3 ISO, while that of using high-pressure carbonation whiteness increased the most and was 71.7 ISO. In this process the color of the modified fly ash changed from dark brown to near white.

7.11 DETERMINING THE EFFECTIVENESS OF FACS AS A RETENTION AID AND FILLER

In this filler study, PCC filled stock was compared to pH-adjusted FACS filled stock. In all studied categories, it was shown that the FACS performed much better at all ash levels. The drainage time decreased, and the air permeability was significantly lowered, as was the turbidity. The study also showed that there appears to be a positive effect when FACS is used in conjunction with PCC.

In this study, 29 experiments were performed to find the nature of FACS filler as a retention aid. In the first ten experiments, a percentage of FACS was added in the range of 0.0 to 1.0%, with a fixed amount of both cationic polymer (0.02%) and precipitated calcium carbonate (7.5%). In control conditions, turbidity and permeability are high as compared to FACS with polymer. The paper air permeability decreased

TABLE 7.5
FACS as a Filler Material and Compared With PCC

Sample	Description	Filler	% Added	Type of Polymer	% Added	Retention Aid	% Added	Drainage (Sec.)	Turbidity (NTU)	Permeability
1	Control	PCC	7.5	Cationic polymer	0.02%	FACS	0.00%	20.01	250	337.43
2	FACS with Polymer	PCC	7.5	Cationic polymer	0.02%		0.00%	17.67	160	325.9
3		PCC	7.5	Cationic polymer	0.02%	FACS	0.03%	17.25	150	325.07
4		PCC	7.5	Cationic polymer	0.02%	FACS	0.05%	17	150	325
5		PCC	7.5	Cationic polymer	0.02%	FACS	0.08%	18.28	150	329.1
6		PCC	7.5	Cationic polymer	0.02%	FACS	0.10%	17.6	150	327.4
7		PCC	7.5	Cationic polymer	0.02%	FACS	0.25%	17.32	155	325
8		PCC	7.5	Cationic polymer	0.02%	FACS	0.50%	17.65	150	325.2
9		PCC	7.5	Cationic polymer	0.02%	FACS	0.75%	17.44	153	316.8
10		PCC	7.5	Cationic polymer	0.02%	FACS	1.00%	17.54	154	327.16
15	FACS without Polymer	PCC	7.5			FACS	0.00%	18.44	182	331.3
16		PCC	7.5			FACS	0.03%	18.02	188	330.02
17		PCC	7.5			FACS	0.05%	18.84	177	331.1
18		PCC	7.5			FACS	0.08%	17.99	186	328
19		PCC	7.5			FACS	0.10%	18.77	180	329.9
20		PCC	7.5			FACS	0.50%	19.12	173	331.93
21		PCC	7.5			FACS	1.00%	18.25	178	329.28
26	FACS treated with Phosporic Acid	PCC	7.5				0.00%	17.85	180	329.68
27		PCC	7.5			FACS	0.05%	17.73	181	330.12
28		PCC	7.5			FACS	0.50%	18.76	187	335.5
29		PCC	7.5			FACS	1.00%	18.28	169	329.08

linearly with its void fraction. When retention aid was added to the papermaking process, it retained fines and filler, resulting in decreased voids and turbidity. The reduction in turbidity is directly related to the colloidal particles present in the water. Re-flocculation on a microscale imparts high permeability to the sheet, which is favorable in both cases for dewatering, press-dryness, and the ability to be dried (Lindstrom 1987).

In papermaking, filler-specific surface areas have a huge effect on permeability. It is also affected by the final mat formation, that is, uniform or non-uniform formation (Han 1969). High and low permeability both can arise depending upon the retention program, that is, dual component (coagulation and flocculation) or triple component (coagulation, flocculation, and micro-particles) and degree of refining (Knauf et al.1986). FACS fillers have aggregation properties, so it is very easy to retain on wire and felt part of the papermaking that why FACS filler have higher FPAR and FPR than the conventional filler like PCC, GCC (Zhang et al. 2013).

FACS fillers have aggregation properties, making them very easy to retain on the wire and felt sections of the paper machine; therefore, FACS fillers have higher FPAR and FPR than conventional fillers like PCC and GCC (Shen et al. 2010a). Similar results were observed in the rest of the cases where FACS filler was used without polymer and with phosphoric acid. From the result, FACS was better than the control (to which no FACS was added as retention aid) condition. However, the pH-adjusted FACS at higher concentrations showed that it could equal the performance of the retention aid in turbidity, while the non–pH-treated FACS reached a similar air permeability.

7.12 DETERMINING THE EFFECT OF FACS FILLER ON PAPER PROPERTIES

As per the literature, with increasing doses of filler, strength properties decease, whereas optical properties increase (Cheng 2009). Cationic copolymer of acrylamide (CPAM) is one of the most commonly used retention aids for improving filler and fine retention (Krochak et al. 2012). Particles with high specific surface area can help in improving fine and filler retention in the final newsprint sheet due to their high adsorption of fibers and fines. Filler particle size, shape, and distribution are important for higher retention (Hubbe et al. 2009). In the present study, higher-pressure carbonation not only provides a whiteness gain, it also modifies the particle morphology that is best suited for retention and drainage.

The opacity of paper is related to the light-scattering coefficient, number of voids, and solid-to-air interface. The higher the number of voids and solid-to-air interface, the higher the light scattering coefficient of the sheet (Klungness et al. 2000). The refining of fibers flattening the fiber result reduction in the number of voids thus affecting the opacity of filled sheet (Chowdhury et al. 2005). Results obtained from

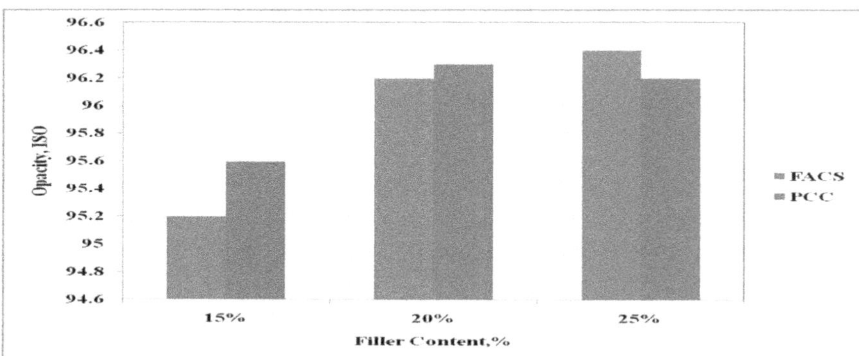

FIGURE 7.4 Effect of filler content on the opacity of paper.

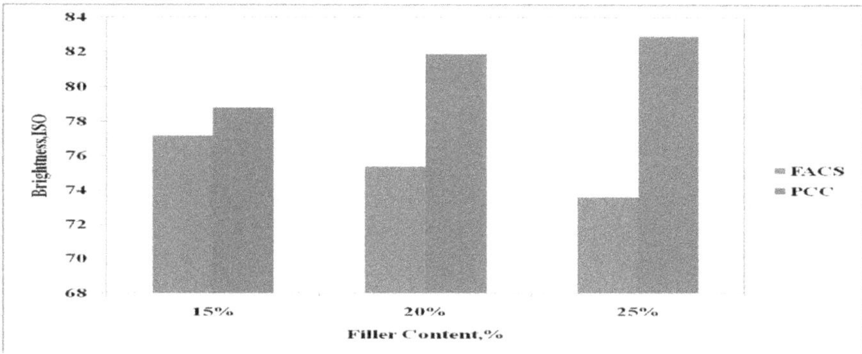

FIGURE 7.5 Effect of filler content on the brightness of paper.

the present study show FACS opacity is better than that of uncontrolled paper sheets and similar to that of PCC-filled writing printing paper in all three cases due to its porous structure, similar particle size distribution, and similar average particle size as compared to PCC filler. Hence, FACS filler overcomes the drawback of different average particle size compared to PCC filler, as noted in the previous study (Song et al. 2012; Verma et al. 2018).

The brightness of different samples of FACS and PCC-filled writing printing paper is shown in Figure 7.5. As the percentage filler loading increases, brightness also increases.

Since PCC has high brightness, PCC-filled sheets have higher brightness as compared to FACS-filled paper and unfilled sheets. However, FACS overcomes the drawback of low brightness of fly ash in various studies (Sinha 2008; Fan and Qian 2012). Results obtained from FACS-filled sheets at different filler levels are encouraging and can be used for the production of any grade of paper.

Paper bulk has a direct relationship with the stiffness of any grade of paper. The higher the bulk, the higher the stiffness of the paper (Chauhan et al. 2012). An increase in bulk means an increase in stiffness, resulting in reduction in the production cost of any grade of paper (Ridgway and Gane 2012). The bulk of the FACS-filled sheet is very high as compared to PCC-filled and base sheets. Unfilled sheet stiffness was 1.5, but when 15 to 25% filler of FACS was used in the base sheet, the stiffness increased remarkably from 1.4 to 2.0 as compared to PCC and unfilled sheets due to higher bulk and strong bonding.

Overall, FACS filler is better than PCC filler for stiffness improvement, as the FACS bulk is very high, as shown in Figure 7.6.

The caliper and smoothness of paper are related to each other. Increasing the filler in paper sheets increases the smoothness of the paper (Lourenco et al. 2014), but it is a very difficult task to maintain caliper as filler increases. In the present study, the smoothness of PCC-filled sheets at all filler percentages was higher than that of FACS-filled sheets and unfilled sheets because PCC filler particles fill the voids within the sheet, resulting in a smoother surface. The results of the smoothness of various filler levels are shown in Figure 7.7.

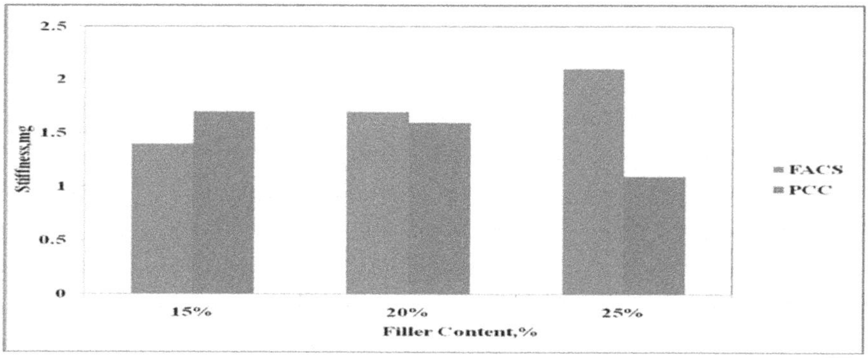

FIGURE 7.6 Effect of filler content on the stiffness of paper.

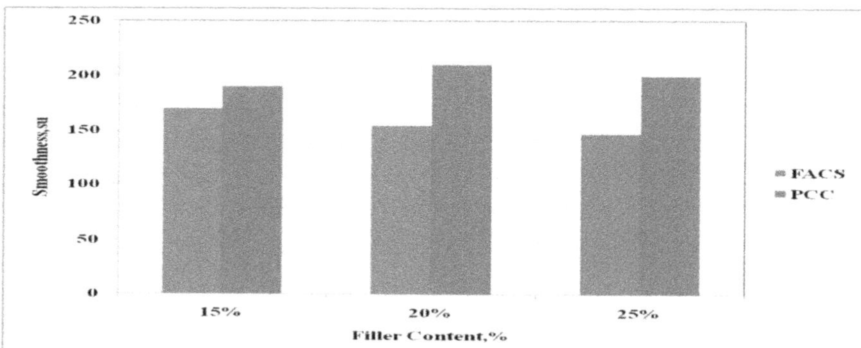

FIGURE 7.7 Effect of filler content on the smoothness of paper.

The FPR (first pass retention) and FPAR (first pass ash retention) of PCC-filled and FACS-filled sheets at various filler percentages are shown in Figure 7.8. Filler aggregation plays an important role for FPR and FPAR in sheets (Cheng 2009), so in this study, fly ash–based calcium silicate was aggregated using starch before integrating into the pulp in order to get the desired result. Aggregation of filler not only helps in FPAR and FPR, it also helps in achieving excellent paper properties, and it is desirable for both economics and the environment, as we used fly ash–based filler for this study (Chauhan 2012). It is evident from our results that one of the most profound impacts is on the increase in first pass retention from 43.5 to 68.9%. It is clear from this data that pre-flocculation processes used in FACS-filled sheets not only control the negative impact on various strength and optical properties of filler-filled sheet, they also help in FPR and FPAR (Sang et al. 2011). The addition of 15%, 20%, and 25% FACS filler to sheets showed a nearly 40% increase in first pass ash retention. An FPAR increase is achieved by good retention aids and preflocculation techniques used in filler treatment before blending with newsprint pulp.

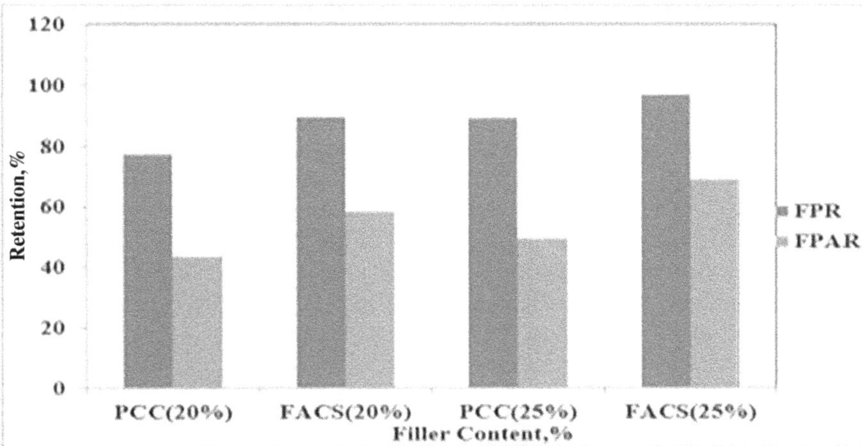

FIGURE 7.8 FPR & FPAR for different filler ratio.

7.13 EFFECT OF PRE-TREATMENT OF FILLER USING PHOSPHORIC ACID AND ATC

The use of phosphoric acid as a modifier for filler is widely recognized among researchers (Peng et al. 1998; 2001; 2003). The optimal dosage of phosphoric acid was determined to be 0.2 wt% relative to the weight of $CaCO_3$ (Pang, 2001). The application of ATC enhances the aggregation of filler and fines within the pulp slurry, leading to improved first pass retention and first pass ash retention. Additionally, ATC reduces steam consumption in the drying section.

a) Comparison of H_3PO_4 for 1 minute versus ATC for 1 minute:

The findings indicate that the blank (unfilled) and carbonate/FACS filled handsheets were adequately sized using the standard chemical dosing rates employed by the local paper mill in Kanpur. The caliper and bulk measurements in both scenarios are identical; however, this pattern does not hold for burst and breaking length. Notably, there is a significant increase in the case of ATC (1 Min), specifically a burst increase of (+8.8%) and a breaking length increase of (+16.6%), although the loss of stiffness (-14.4%) is too severe with ATC. The optical properties and other remaining characteristics are consistent across both scenarios.

b) Comparison of H_3PO_4 for 1 minute versus H_3PO_4 for 30 minute:

The caliper and bulk measurements in both cases (1 min vs 30 mins) are the same, yet this trend does not apply to burst and breaking length, which significantly increases in the case of H_3PO_4 for 30 Min (with a burst increase of 3.5% in Kpa, while the breaking length remains unchanged). Furthermore, a stiffness gain of (+4%) is observed when H_3PO_4 is added after 30 min. All other properties remain nearly identical.

7.14 EFFECT OF PRE-TREATMENT OF FILLER USING STARCH

The aim for this study was to check the high-percentage loading effect on strength properties. Various approaches have been investigated to address the challenges of increasing ash content in sheets. In the current study, the concept of pre-flocculation of fillers with the help of cationic starch, ATC, phosphoric acid, and its particle size distribution using Malvern Mastersizer 2000 prior to their addition to the wet end has been analyzed. The term pre-flocculated means the modification of filler particles in agglomerates through treatment with starch before adding to fiber. This balanced modification is expected to lead to a reduced negative impact of filler addition on paper strength (Esser et al., 2009). Due to high-speed paper machines, filler size and flocs should be stable and shear resistant (Cheng et al. 2011).

A study was conducted for paper grades of 50–57 and 60 gsm. The furnish was a mix of NBKP, LBKP (fibria and self-made), and broke. The starting base paper ash was 15.8%. Acid-resistant properties of fillers under control conditions can be achieved with the help of starch derivatives. Cationic and anionic starch should be used to coat fillers to agglomerate them so that they cannot pass through the wire section (Lambert and Lowes 1975). Pre-flocculation of filler using starch derivatives has improved strength properties (Harvey and Klem 1989). A starch–soap complex was also used as a coating additive for filler, which enhanced strength as well as sizing efficiency (Kurrle 1996). PCC modified using starch mixing, grinding, dewatering, and aggregation enhances the tear, tensile, and folding strength of paper (Zhao et al. 2005). Modification of clay using starch and ammonium sulfate showed greater benefits in paper strength properties (Yoon and Deng 2006).

In this study, starch was heated at 3.0% solids for at 160°C (inlet temperature) and 40°C (outlet temperature), mixed with 10% solid FACS slurry, and stirred at 220 rpm. The treated filler particle size and its distribution were measured by using a Malvern Mastersizer 2000 instrument.

The pre-flocculation step is continuously monitored, and the strong flocs produced are frequently measured to achieve better control of filler particle size. Initially, the treated filler was added to the process to maintain the current base ash level of 15.5 for 50 gsm, and then filler dosing rates were increased to raise the base sheet ash level up to 18.5. A similar process was followed for 54, 57, and 60 gsm paper with a current base ash level of 17, 16.7, and 17.5 to raise base ash level up to 20.2, 19.2, and 20.9.

It was found that modified FACS exhibits changes with respect to morphology and particle size when compared to unmodified FACS, and the particle size of treated FACS was much larger than that of the untreated FACS. The treated filler floc enables a size that covers less fiber surface when compared to traditionally added untreated filler, thus increasing the surface area for bonding, needed to preserve the sheet strength, that is, minimizing the interference with the fiber-to-fiber bonding strength network (Gaudreault et al. 2009).

Summary of the properties of sheets with untreated filler with treated filler is given below in Table

TABLE 7.6A
Summarizes the Properties of Sheets with Untreated Filler with Treated Filler

Parameters	UNIT	50 gsm		54gsm		57gsm		60gsm	
		Untreated Filler	Treated Filler	Untreated Filler	Treated Filler	Untreated Filler	Treated Filler	Untreated Filler	Treated Filler
Actual GSM	g/m^2	50.2	50.1	54.57.20	54.1	57.2	57.1	60	59.9
Thickness	^m	70	70	75	75	80	79	83	83
Bulk	Cm^3/g	1.39	1.4	1.39	1.41	1.4	1.39	1.38	1.38
Stiffness,MD		0.84	0.8	0.97	0.9	1.22	1.15	1.3	1.21
Stiffness,CD		0.32	0.25	0.37	0.32	0.44	0.39	0.5	0.45
Bendt Smoothness,T	ml/min	153	162	152	179	182	170	166	181
Gurley Porosity	Sec/100 ml	28	27	28	25	27	24	27	26
Wax pick,T		14	14	14	14	14	14	14	14
Wax Pick,W		14	14	14	14	14	14	14	14
Paper Ash	%	15.5	18.5	17	20.2	16.7	19.2	17.5	20.9
Tear Strength,CD	gm	34	33	36	35	39	38	39	37
Tensile Strength,MD	Kg/15 mm	4.2	3.71	4.37	4.06	4.53	4.09	4.53	4.2
Breaking Length,MD	mts	5582	5200	5401	4765	5300	5600	5050	4700

TABLE 7.6B
Ash Gain During Various Grammage Paper

Paper grade	Untreated FACS	Treated FACS	% Ash Gain
50 gsm	15.5	18.5	+3.0
54gsm	17	20.2	+3.2
57gsm	16.7	19.2	+2.5
60 gsm	17.5	20.9	+3.4

Treated fillers particle size and particle size distribution have been checked by Malvern Instrument. Paper sheets collected and measured on physical and structural properties using the corresponding ISO standards.

7.15 CONCLUSION

Based on the results, it appears that FACS is not confirmed as a retention aid at very high dosages, but at low dosages, it can be used as a retention aid. However, there is evidence that FACS has the potential to be used as a primary or secondary filler in paper stock. First pass retention and first pass ash retention are very high compared to conventional fillers like PCC, talc, and GCC under identical conditions with and without retention aids. FACS filler has a high sizing demand as compared to conventional filler; on increasing the sizing dosages, FACS filler sheets can be properly sized, but this requires using greater amounts of sizing chemicals. A blend of FACS with conventional filler can be a solution to this problem. Low-cost waste fly ash has been used as a coating pigment in newsprint for quality improvement, which provides economic and environmental benefits. The most significant increase was in brightness, which generates a potential for higher revenue in newsprint. Newsprint made with a calcium silicate pigment coating can compete with imported newsprint. Fly ash generated from a power plant in a newsprint mill can be used as a potential coating pigment in the size press for energy savings during drying and results in a higher opacity of newsprint and superior mechanical strength properties. The production of modified fly ash–based coating pigment using the high-pressure carbonation process shows excellent optical and strength properties when compared to PCC and the base newsprint sheet.

Newsprint sheet strength increases, contrary to various literature data stating that increasing ash decreases the sheet strength. The addition of filler also offers the potential to replace more expensive fiber components, thus reducing the cost of manufacturing. The pre-flocculation method can help in increasing ash in the final newsprint sheet as well as balancing the various strength properties of newsprint. Newsprint sheet brightness increased with the addition of fly ash-based calcium silicate filler; increasing brightness can result in a higher newsprint selling price. One of the most profound findings of this research was an increase in first pass ash retention from 30% to 70%. Using fly ash–based calcium silicate filler in newsprint will allow locally produced newsprint to compete with expensive imported paper as well as allowing newsprint to be more competitive in international markets. Fly ash–based filler has unique properties compared to PCC due to high-pressure carbonation and pre-flocculation techniques. Modified FACS has high FPR and FPAR as compared to unmodified PCC with low starch consumption. Modified FACS shows a slight drop in opacity and brightness as compared to unmodified PCC, as treated fillers have a much less negative impact on optical properties. Bulk reduction is a direct consequence of higher filler loading, as reported in various studies, but pre-flocculation techniques have a low or negligible effect on bulk.

Filler in paper comes from two sources: fillers from a direct supply tank and from broke furnish. Therefore, it is quite obvious that we must carefully manage the filler coming from the supply tank in order to increase the ash level in the base sheet, since we are treating only that filler. Pre-defined filler particle size, floc size/shear resisting power of flocs, and controlled particle size distribution can reduce the effects of fillers on sheet strength, so when increasing ash content in paper, the span should be lower. It also appears that pre-flocculated filler decreases sheet bulk less than untreated fine particle filler, probably because the former has higher porosity and thus lower density than the latter. Bulk can also be preserved by reducing calendaring load. Sizing chemical dosage was kept slightly on the higher side to maintain the desired Cobb value. With this dosage, hard sizing was observed; therefore, there is margin to optimize the sizing chemical dosage. A slight decrease was observed in paper strength properties like breaking length, burst factor, and double fold.

BIBLIOGRAPHY

Hubbe, M. A., & Gill, R. A. (2016). Fillers for papermaking: A review of their properties, usage practices, and mechanistic role. *BioResources*, *11*(1), 2886–2963.

Alava, M., & Niskanen, K. (2006). The physics of paper. *Reports on Progress in Physics*, *69*(3), 669–723.

Shen, J., Song, Z. Q., Qian, X. R., & Liu, W. (2009). Modification of papermaking grade fillers: A review. *BioResources*, *4*(3), 1190–1209.

Chen, H., Chen, K., Yang, R., Yang, F., & Gao, W. (2011). Use of aluminum trihydrate filler to improve strength properties of cellulosic paper. *BioResources*, *6*(3), 2399–2410.

Chen, X., Qian, X., & An, X. (2011). Calcium carbonate whiskers as papermaking filler. *BioResources*, *6*(3), 2435–2447.

Gullichsen, J. (2000). Fiber line operations. In J. Gullichsen & C.-J. Fogelholm (Eds.), *Chemical pulping: Papermaking science and technology* (Book 6A). Fapet Oy.

Solomon, D. H., & Hawthorne, D. G. (1983). *Chemistry of pigments and fillers*. Wiley.

Mathur, R. M., Thapliyal, B. P., et al. (2009). Challenges confronting Indian paper industry in a changing scenario. *Indian Pulp and Paper Technical Association Journal*, *21*(3), 96.

Jain, R. K., Eriksen, Ø., Isaksson, A., Kumar, V., & Marwaha, A. (2017). Development of appropriate technologies for pulp and paper sector. In *PaperEx South India Conference Proceedings*. Chennai, India.

Gallagher, T. M. (1990). *Retention: The key to efficient papermaking*. Nalco Chemical Company.

Norell, M., Johansson, K., & Persson, M. (1999). Retention and drainage. In L. Neimo (Ed.), *Papermaking Chemistry* (pp. 42–81).

Cadotte, M., Tellier, M. E., Blanco, A., Fuente, E., van de Ven, T. G. M., & Paris, J. (2007). Flocculation, retention, and drainage in papermaking. *Canadian Journal of Chemical Engineering*, *85*, 240–248.

Gregory, J. (1985). Polymeric flocculants in flocculation and sedimentation. In *Engineering Foundation Conference Proceedings* (pp. 125–137).

Berlin, A. A., & Kislenko, V. N. (1995). Kinetic model of suspension flocculation by polymers. *Colloids and Surfaces, 104*, 67–72.
Biggs, S., Habgood, M., Jameson, G. J., & Yan, Y. (2000). Aggregate structures in flocculation. *Chemical Engineering Journal, 80*, 13–22.
Bremmell, K. E., Jameson, G. J., & Biggs, S. (1998). Polyelectrolyte adsorption kinetics. *Colloids and Surfaces A, 139*, 199–211.
Blanco, A., Fuente, E., Negro, C., & Tijero, J. (2002). Flocculation monitoring using FBRM. *Canadian Journal of Chemical Engineering, 80*, 734–740.
Eklund, D., & Lindström, T. (1991). *Paper chemistry: An introduction.* DT Paper Science Publications.
Clémençon, I., & Geril, A. (1999). Flocculant/microparticle retention systems. *Nordic Pulp and Paper Research Journal, 14*(1), 23–29.
Wågberg, L., Åström, I., & Swerin, A. (1996). Flocculation mechanisms in papermaking. *TAPPI Journal, 79*(6), 157–164.
Swerin, A., Sjödin, U., & Ödberg, L. (1993). Flocculation of fiber suspensions. *Nordic Pulp and Paper Research Journal*, 389–399.
Lindström, T., & Söderberg, G. (1986). Mechanism of AKD sizing. *Nordic Pulp and Paper Research Journal, 1*(2), 31.
Isogai, A., Taniguchi, R., Onabe, F., & Usuda, M. (1992). Sizing mechanism of AKD. *Nordic Pulp and Paper Research Journal, 7*(4), 193.
Champ, S., & Ettl, R. (2004). AKD retention dynamics. *Journal of Pulp and Paper Science, 30*(12), 322.
Johansson, J., & Lindström, T. (2004). AKD sizing efficiency. *Nordic Pulp and Paper Research Journal, 19*(3), 330.
Ravnjak, D., Plazl, I., & Moze, A. (2007). AKD particle deposition kinetics. *Colloid and Polymer Science, 285*, 907.
Yuan, J. H. (2008). *Fiber loading technology for high ash paper* (PhD thesis). Nanjing Forestry University.
Lu, Z. H., Li, M., & Liu, H. J. (2010). Introduction to paper fillers. *Journal of Hubei Papermaking*, 40–44.
Han, C. (2009). *Starch-coated PCC in papermaking* (Master's thesis). Nanjing Forestry University.
Deng, Y., Yoon, S. Y., Ragauskas, A. J., & White, D. (2008). Papermaking methods and compositions. *U.S. Patent.*
Yoon, S. Y., & Deng, Y. (2007). Strength properties of composite-filled paper. *Industrial & Engineering Chemistry Research, 46*, 4883–4890.
Zhao, Y., Kim, D., White, D., Deng, Y., & Ragauskas, A. J. (2008). New paradigm for linerboard fillers. *TAPPI Journal, 7*(3), 3–5.
Yan, Z., Liu, Q., & Deng, Y. (2005). Paper strength improvement using modified clay. *Journal of Applied Polymer Science, 97*, 44–50.
Cheng, W., Broadus, K., & Ancona, M. (2011). Increased filler use and fiber savings. In *TAPPI PaperCon Proceedings.*
Scott, W. E. (1996). *Principles of wet end chemistry.* TAPPI Press.
Seppänen, R., Ström, G., & Elftonson, J. (2000). Heteroflocculation of kaolin. *Colloids and Surfaces A*, 131–141.
Blanco, A., Negro, C., Fuente, E., & Tijero, J. (2005). Effect of shear on flocculation. *Industrial & Engineering Chemistry Research, 44*, 9105–9112.
Sharma, A., et al. (2012). ASA sizing efficacy with fillers. *IPPTA Journal, 24*(3), 93–98.
Biricik, Y., Sonmez, S., & Ozden, O. (2011). Surface sizing effects. *Asian Journal of Chemistry, 23*(7), 3151.
Lindström, T., & Florén, T. (1987). Effect of filler particle size. *Nordic Pulp and Paper Research Journal, 2*(4), 142–145.

Han, S. T. (1969). Compressibility of fiber mats. *Pulp and Paper Magazine of Canada*, *70*(9), T134–T146.
Knauf, G. H., & Doshi, M. R. (1986). Aerodynamic porosity calculations. In *TAPPI Conference Proceedings*.
Hubbe, M. A., Nanko, H., & McNeal, M. (2009). Retention aid interactions. *BioResources*, *4*(2), 850–906.
Klungness, J., et al. (2000). Lightweight high-opacity paper. *Nordic Pulp and Paper Research Journal*, *15*(5), 345–350.
Chowdhury, P., et al. (2005). Improving paper opacity. *IPPTA Journal*, *17*(3), 57.
Verma, A. K., Suresh, S., & Mohta, D. (2018). Fly ash for newsprint improvement. *BioResources*, *13*(3), 5765–5777.
Ridgway, C. J., & Gane, P. A. (2012). NFC-pigment composites. *Cellulose*, *19*(2), 547–560.
Sang, Y., McQuaid, M., & Englezos, P. (2011). Pre-flocculation of PCC filler. *BioResources*, *7*(1), 354–373.
Gaudreault, R., Di Cesare, N., Weitz, D. A., & van de Ven, T. G. M. (2009). Flocculation kinetics of PCC. *Colloids and Surfaces A*, *340*, 56–65.

Index

For Product Safety Concerns and Information please contact our EU
representative GPSR@taylorandfrancis.com
Taylor & Francis Verlag GmbH, Kaufingerstraße 24, 80331 München, Germany

www.ingramcontent.com/pod-product-compliance
Ingram Content Group UK Ltd.
Pitfield, Milton Keynes, MK11 3LW, UK
UKHW020208200926
475702UK00004B/665
* 9 7 8 1 0 3 2 5 8 9 2 6 8 *